SPITE

The Upside of Your Dark Side

SPITE

Simon McCarthy-Jones

BASIC BOOKS

New York

Basic Books
Hachette Book Group
1290 Avenue of the Americas, New York, NY 10104
www.basicbooks.com

Printed in the United States of America

Originally published in 2020 by Oneworld Publications in Great Britain

First US Edition: April 2021

Published by Basic Books, an imprint of Perseus Books, LLC, a subsidiary of Hachette Book Group, Inc. The Basic Books name and logo is a trademark of the Hachette Book Group.

The Hachette Speakers Bureau provides a wide range of authors for speaking events. To find out more, go to www.hachettespeakersbureau.com or call (866) 376-6591.

The publisher is not responsible for websites (or their content) that are not owned by the publisher.

Print book interior design by Amy Quinn.

Library of Congress Cataloging-in-Publication Data
Names: McCarthy-Jones, Simon, 1978– author.
Title: Spite: the upside of your dark side | Simon McCarthy-Jones.
Description: First US edition. | New York: Basic Books, [2021] | Includes bibliographical references and index.
Identifiers: LCCN 2020044414 | ISBN 9781541646995 (hardcover) | ISBN 9781541646988 (ebook)
Subjects: LCSH: Revenge—Social aspects.
Classification: LCC BF637.R48 M377 2021 | DDC 152.4—dc23
LC record available at https://lccn.loc.gov/2020044414

ISBNs: 978-1-5416-4699-5 (hardcover); 978-1-5416-4698-8 (ebook)

LSC-C

Printing 1, 2021

CONTENTS

Introduction THE FOURTH BEHAVIOUR 1

One ULTIMATUMS 9

Two COUNTERDOMINANT SPITE 39

Three DOMINANT SPITE 79

Four SPITE, EVOLUTION, AND PUNISHMENT 93

Five SPITE AND FREEDOM 111

Six SPITE AND POLITICS 135

Seven SPITE AND THE SACRED 163

Eight THE FUTURE OF SPITE 189

ACKNOWLEDGMENTS 209

NOTES 211

INDEX 249

Introduction
THE FOURTH BEHAVIOR

Spite runs deep. We find it in our oldest stories. It is there in the myths of ancient Greece. Medea kills her children, just to spite her unfaithful husband, Jason. Achilles refuses to help his Greek comrades fight because one of them has stolen his slave. Folklore tells of spite. A magical being offers to grant a man one wish. Naturally, there is a catch. Whatever he gets, his hated neighbor will get double. The man wishes to be blind in one eye.[1] Such stories, although buried in time, still speak of an instantly recognizable behavior.

Today, we know that spite can be petty. A driver lingers in a parking space, just to make you wait. A neighbor puts up a fence, solely to block your view. We may also realize how damaging spite can be. A spouse seeks custody of a child, just to get back at their ex. A voter supports a candidate they hope will cause chaos. But are we prepared to recognize that spite may have a positive side?

What exactly is spite? According to the American psychologist David Marcus, a spiteful act is one where you harm another person and harm yourself in the process.[2] This is a "strong" definition of spite. In weaker definitions, spite is harming another while only risking harm to yourself. It can also be harming another while not personally benefiting from doing so.[3] Yet, as Marcus points out, a strong definition of spite, in which harming another entails a personal cost, helps differentiate it from other hostile or sadistic behaviors.

Indeed, a helpful way to understand spite is to look at what it isn't. When we consider the costs and benefits of our actions, there are four basic ways we can interact with another person.[4] Two behaviors involve direct perks for us. We can act in a way that benefits both ourselves and the other (cooperation) or in a way that benefits ourselves but not the other (selfishness). A third behavior involves a cost to us but a benefit to the other. This is altruism. Researchers have dedicated lifetimes to the study of cooperation, selfishness, and altruism. But there is a fourth behavior, spite. Here we behave in a way that harms both ourselves and the other. This behavior has been left in the shadows, which is not a safe place for it to be. We need to shine a light on spite.

Spite is challenging to explain. It seems to present an evolutionary puzzle. Why would natural selection not have weeded out a behavior in which everybody loses? Spite should never have survived. If your spite benefits you in the long run, then its continued existence becomes comprehensible. But what about spiteful acts that don't give you long-term benefits? How can we explain those? Do such acts even exist?

Spite also poses a problem for economists. What kind of person acts against their self-interest? For the longest time, economists didn't think there was a problem to explain. The famous eighteenth-century economist Adam Smith claimed that people were "not very frequently under the influence" of spite, and that even if it did occur we would be "restrained by prudential considerations."[5] Much later, in the 1970s, the American economist Gordon Tullock claimed that the average human was about 95 percent selfish.[6] In the "greed is good" era of the 1980s, many may have believed that this estimate was on the low side.

Economists viewed humans as a creature called *Homo economicus*—a being that acted rationally to maximize its self-interest. Self-interest was typically, though not always, understood in financial terms.[7] Yet, as I will discuss in Chapter 1, back in 1977 a groundbreaking study found that people were often quite happy to turn down free money. Adam Smith had been overoptimistic. Something very real and very powerful lurked in Tullock's residual percentage.

Spite involves harm, but what constitutes harm? Who gets to decide whether an act is harmful and thereby has the power to define an act as spiteful? To take an extreme example, does a suicide bomber, who thinks they will be rewarded in the next life and their family compensated in this life, harm themselves or not?

Evolutionary biologists possess an objective measure of harm: a loss of fitness (reproductive success). We will look at spiteful acts involving a loss of personal fitness, so-called evolutionary spite, in Chapter 4. In contrast, economists and psychologists tend to focus on harm in the form of immediate personal costs. This "psychological spite" can turn out to have unforeseen long-term personal benefits. Such spite ages well, maturing into selfishness.

Once we are happy with our definition of spite, two questions remain. First, what drives someone to act spitefully in the moment? That is, how does spite work? This is spite's proximate explanation. Second, what is the deeper reason we are spiteful? Why does spite exist? What is its evolutionary function? This is the ultimate explanation of spite. To take an example from another area: why do babies cry? The proximate explanation may be cold or hunger, but the ultimate explanation is to get care from its parents.[8] What are the equivalent answers for spite?

When we have an ultimate explanation of spite, we can begin to consider the pressing question of how spite shapes the modern world. A love of sugar and fat helped our ancestors, pushing them to eat high-energy foods. Yet in the Western world today, where cheap sugary and fatty foods are ubiquitous, what was once adaptive now causes diabetes and heart disease. What happens when our evolved spiteful side runs into a world it was never meant to deal with? What are the effects of spite in a world with levels of economic inequality, perceived injustice, and social media–enabled communication that would be utterly alien to our ancestors?

The problem is pressing because spite seems more than dangerous. From some angles it looks like human kryptonite. Spite is, by definition, the exact opposite of cooperation, which is worrying, because cooperation is our species' superpower. Our success as a species has come from our remarkable ability to work together. Although even slime molds cooperate,[9] we turn cooperation up to eleven. This allows us to live in large groups with nonrelatives, something that our less cooperative primate cousins cannot do.[10] Our ability to cooperate at scale keeps us safe from a *Planet of the Apes*–style takeover for the time being, yet there are many other things to worry about short of primate overlords. If spite damages cooperation, it would not only hamper human progress; it could also reduce our ability to solve the complex global problems we face.[11] The world is getting better, but progress is not guaranteed.[12]

Spite can also be terrifying. Is there anything more frightening than an adversary unfettered by the bonds of self-interest? Selfishness can be a problem. But at least we can reason with a selfish person's self-interest. What do you say to a spiteful person who values your suffering more than their own well-being? They are like a Terminator. They can't be bargained with, can't be reasoned with, and absolutely will not stop, ever, until you are, if not dead, at least inconvenienced. Unfortunately, such creatures are not limited to science fiction.

Late in the Second World War, Germany's fight against Russia was intensifying. Hitler had to balance using trains to transport Jews to their death in the east against using the same trains to carry vital weapons, fuel, and supplies to his army fighting the Russians.[13] Hitler chose Holocaust. He was prepared to risk the destruction of Germany to eliminate the Jewish people. The horror of spite has no bounds.

Spite poses clear and horrific dangers. We need to understand it in order to control it. To do this, we are forced to look closely at spite. And when we do, something else emerges. What we find forces us to reconsider whether we may have misunderstood spite. The American philosopher John Rawls argued that moral virtues are character traits that we should rationally want other people to have.[14] Spite, he claimed, is not something we should want others to have, and therefore it is a vice that is "to everyone's detriment." But is it?

It turns out that spite can be a force for good. It can help us excel. It can help us create. And it does not necessarily threaten cooperation. In fact, paradoxically, it may spur it. Spite does not inevitably produce injustice. In fact, it may be one of our most powerful tools for preventing it. As long as injustice and inequitable inequality persist, we may need spite.

The book of Ezekiel tells of a vision the biblical prophet experienced when he was thirty years old. A great storm came from the north. Inside was a fire. From the fire came a four-sided creature. Each side had a different face: one human, one lion, one ox, and another an eagle. Human nature, like the creature of Ezekiel's vision, is chimerical. We, too, show the world four faces: selfishness, altruism, cooperation, and spite. We are multifaceted, being neither angels nor demons. To understand ourselves, we need to know all our parts, not merely one side. We are an adaptable ape with a unique repertoire of behaviors. Which of these we deploy, to reap their benefits, depends on the circumstances we face. Spitefulness isn't a dark stain on our soul; it is part of our soul. Like the creature of Ezekiel's vision, our sides are interconnected. It is not simply that we have dark and light sides. Our dark side may create the light. We need to be prepared to look for the origins of virtue in vice.

One
ULTIMATUMS

he West German autumn of 1977 was one of ultimatums.
A new generation of Germans, born after the horrors of
the Second World War, had reached adulthood. These
were the *Nachgeborenen*, the "later-born." Though raised in
the shadows of their parents' sins, they were not culpable
for the crimes of the Third Reich. Still, they felt tainted by
them. When they saw remnants of Nazism in their state,
they were appalled. For some, resistance had to go beyond
words. How, they asked, could you reason with the people
who had created Auschwitz? Such sentiments gave birth to
the Baader-Meinhof group: a radical left-wing organization

that aimed to overthrow capitalism, imperialism, and fascism. In the end, it merely produced what always comes from stirring Marxism with machine guns in the cauldron of youth: blood.

On the evening of September 5, 1977, members of Baader-Meinhof wheeled a blue baby stroller into the middle of a quiet road in Cologne. A large Mercedes came around the corner. In the backseat was Hanns-Martin Schleyer. He had been an officer in the SS during the war. Now he was one of West Germany's most powerful industrialists. Seeing the stroller, Schleyer's driver braked sharply. Their police escort slammed into the back of their car. Baader-Meinhof pounced.

One member, armed with a Heckler and Koch semi-automatic rifle, climbed onto the hood of the police car. He emptied the gun into the men and the machine at his feet. When the shooting was over, three policemen and Schleyer's driver were dead. But Schleyer was alive. Baader-Meinhof took him and issued an ultimatum to the West German government. Several first-generation members of Baader-Meinhof were currently languishing in prison, serving life sentences. Release them, the group demanded, or we will execute Schleyer. As the government stalled for time, the situation worsened.

On October 13, five weeks after Schleyer's kidnapping, the Popular Front for the Liberation of Palestine hijacked Lufthansa flight 181. The front had close ties with Baader-Meinhof. Members of the two groups had crawled through the sands of Jordan together in Palestinian training camps. Eventually, the hijackers forced the pilots to land in Mogadishu, Somalia. Then came the hijackers' ultimatum: meet our demands, including the release of our Baader-Meinhof comrades; otherwise the eighty-plus passengers, the flight crew, and Schleyer will die.

On October 18, events unfolded at pace. The West German government had dispatched the elite GSG 9 commando unit to Somalia. It had formed the unit in the wake of the murder of Israeli athletes at the 1972 Munich Summer Olympics. At two o'clock in the morning, Somali soldiers lit a fire on the tarmac in front of the plane. As the hijackers entered the cockpit to see what was happening, GSG 9 commandos stormed the plane. In seven minutes they had neutralized the hijackers and rescued all the hostages, alive.

At seven o'clock that same morning, in Stuttgart, Germany, the prison guards of Stammheim Prison began their rounds. News had filtered back to the imprisoned Baader-Meinhof members that their comrades' ultimatums had been unsuccessful. The first member of Baader-Meinhof that the guards checked on was Jan-Carl Raspe. They found him sitting in his cell with a bullet wound to the head. He would die hours later. Suspecting a coordinated series of events, the guards ran to the nearby cell of Andreas Baader.

Baader was a founding member of the group that bore his name. In his younger days he was not a man who read Marx. He was not a man who read Mao. It is unclear if he was a man who read at all.[1] He preferred driving fast cars and sleeping around. His political analysis of the West German state was that it was a "shat-in shithouse."[2] During his imprisonment, the famous French philosopher Jean-Paul Sartre visited him. "Jerk" was Sartre's verdict. Baader had never dealt well with authority. As a child, his mother had taken him boating on a lake and warned him to take care. Baader jumped in the lake. As an adolescent, he feigned lung cancer to get sympathy. As an adult, he bombed and he burnt and he murdered.

In prison, the Baader-Meinhof members had adopted code names from *Moby Dick*. Baader was Ahab. As the guards entered Baader's cell that October morning, they saw Ahab on his back, dead, his head resting in a sea of blood. A later examination found that he had died of a gunshot to the back of the neck.

The guards next ran to the cell of Baader's imprisoned lover, Gudrun Ensslin. She was a pastor's daughter, but this lady's cheek was not for turning. For her, violence was the only answer to violence. Through the darkness, Ensslin appeared to be standing, staring out the window. When the guards looked closer, they saw that her feet were a foot off the ground. She had died by hanging. A fourth member of the group, Irmgard Möller, was found in her bed. She had multiple stab wounds to the chest but would survive.

When news of these deaths reached the Baader-Meinhof members who were holding Schleyer, they shot him in the head. Police found his body the next day, stuffed into the trunk of a green Audi. So ended the German autumn of 1977.

Multiple theories have been put forward for how the imprisoned Baader-Meinhof members died.[3] The official explanation is that their wounds were self-inflicted as part of a suicide pact. It has also been suggested that they committed suicide in a way that would make it appear that the state had murdered them. The aim of such a maneuver would have been to make the government appear tyrannical and inspire its overthrow.

If true, then Baader's codename of Ahab was appropriate. In Herman Melville's novel, Captain Ahab was so consumed with destroying the eponymous white whale that he was prepared to obliterate himself, his ship, and his crew in the process. If the imprisoned Baader-Meinhof members did commit suicide to harm the West German government, there is a word for such behavior. It is the same word we can use to describe Ahab's actions. That word is "spite."

By a remarkable coincidence, the foundations for the study of spite were also laid in Germany in the autumn of 1977.[4] It even happened in the very same city where Baader-Meinhof abducted Schleyer. The design of the research would echo the events of that autumn, focusing on how people responded to ultimatums. The world would be introduced to the Ultimatum Game, which would change our understanding of what kind of creature we are. Four decades on, the results of this game still offer insights into the chimeric messiness of human nature.

SPITE PERMEATES OUR EVERYDAY EXPERIENCE. Put yourself in the situations below. How many apply to you?

1. If I were getting ready to pull out of a space in a busy parking lot and saw someone else waiting impatiently to pull in, I'd take my time just to make them wait.

2. I'd vote for a politician I disliked, just to keep someone else out of office, even if I thought that politician would hurt me and/or my country.

3. If a parent told me not to go out looking messy, I'd make myself look worse, even if it embarrassed me in front of my friends.

4. I'd be prepared to feel cold in my house if it meant others in the house were made to feel uncomfortably cold too.

5. I'd volunteer to stay late at work if it meant my colleagues had to stay late too.

6. If someone were driving close behind me, I'd tap my brakes just to give them a fright, even though doing so would put me in danger.

7. If I were angry at my partner, I'd burn the dinner, even though it meant I would go hungry too.

8. I'd put up something ugly in my yard if I thought it would annoy my neighbors.

9. I'd like to see one of my colleagues fail, even if it means I get a smaller bonus this year.

10. I'd take my time at the checkout line, just to delay someone behind me.

In 2014, David Marcus at Washington State University posed these types of questions to the public. Remarkably, it was the first time anyone had attempted to use a questionnaire to quantify how spiteful we are. For each item, they found that 5 to 10 percent of people agreed with the statement.[5]

Of course, questionnaires only assess how people *say* they would act. People might behave in a very different way when confronted with such situations. This is a long-standing concern in psychological research. Back in the 1930s, a researcher at Stanford University, Richard LaPiere, observed that people would often talk in a racist way but would not act this way when they met someone of a different race. Maybe you can think of an elderly relative who comes out with racist sentiments at home that make you wince, yet who is affable when interacting with people of other races. LaPiere thought this tendency might be the rule rather than the exception. He was able to test his hypothesis when he spent two years traveling around America with a young Chinese couple.

At the time, Chinese people in the United States faced widespread prejudice. When LaPiere and the couple arrived at a hotel or a restaurant, he would fiddle with the luggage while the couple went in ahead of him. LaPiere watched how the proprietors behaved. During ten thousand miles of driving across America, they were not refused service in any of the 184 restaurants they visited. Only one of the sixty-six hotels they visited turned them away. In this sole refusal, the owner proclaimed, "I don't take Japs." Six months later, LaPiere sent questionnaires to all these establishments asking if they would accept or serve Chinese guests. Over 90 percent said they would not.[6] What people say they will do can be very different from what they actually do, which, in this case, we should be thankful for.

Despite this fact, a study of spiteful behavior found similar results to those of Marcus's questionnaire research. The study took advantage of how auctions work. Imagine you see someone leading an auction on eBay for a flat-screen TV with a bid of $50. You suspect the other bidder is prepared to pay at least $200. So even though you have no interest in the TV, you pop in a bid of $100 just to force them to bid more to get it. Sure enough, the other person responds with a bid of $110. You muse for a bit. Can you make them pay more? You decide to bid $150. The other person does nothing. Your heart starts to pump harder. Sweat beads. *Uh-oh*, you think, *I might win this*. Then a bid of $160 pops up from the other person. You sink back in your seat, breathe again, and decide to push your luck no further. You've acted spitefully and gotten away with it. *Why the hell*, you wonder, *did I ever get involved in that? Why did I waste my time, and nearly a lot of money, just so some person I don't know and will never meet won't gain some tenuous advantage over me?* And in case you are wondering, no, this isn't an example from my own life. How dare you. I was bidding on a vase.

Unflattering personal anecdotes aside, in 2012 the economists Erik Kimbrough and J. Philipp Reiss studied the frequency of spiteful bidding in auctions. They set up an auction so that the price the winning bidder would pay was the bid made by the second-highest bidder. Their setup then gave the second-highest bidder the option to spitefully push up the amount the winning bidder would pay, but without risking winning the auction themselves.

The researchers found that a third of people made the most spiteful bid they could most of the time.[7] This is a remarkable amount of spite. However, another third of people spent most of their time making bids that were not at all spiteful. The majority of people were either not spiteful at all or maxed out on spite. Not only does spite appear to be concentrated in certain people; it tends to be an all-or-nothing phenomenon. This is why I believe you if you say that the spiteful scenarios I outlined above are entirely foreign to you. Of course, it also makes me suspect that you are a bit of a do-gooder. In Chapter 2 we will come to the reasons why I'm tempted to slander your decency in this way.

In this auction study, people didn't really incur any personal costs to inflict harm on the other person. At most, they wasted their time and energy to figure out how to increase their bid. Either way, spite was cheap. What would happen if people had to pay a small amount to make a spiteful bid? If it cost people cold, hard cash to be spiteful, would spite vanish?

To answer this, we can turn to another game in which you can win real money. Here, if you choose to act spitefully it comes at a financial cost. This is the Ultimatum Game, created in Cologne during the German autumn of 1977. This simple game has been remarkably influential. In academia, if one other person takes an interest in your work, you are delighted. Over five thousand works have cited the original Ultimatum Game study. It is still used today to research spite. Whether it be in a laboratory in America or a clearing in Borneo, someone somewhere is probably playing it right now.

━━━

AMIDST THE VIOLENCE OF THE German autumn of 1977, thirty-three-year-old Werner Güth arrived to take up a new professorial position at Cologne University. Game theorists were keen for him to study how people behaved, and they gave him 1,000 deutsche marks to do it. The game Güth devised had its roots in his childhood.[8] When he and his brother had to share a slice of cake, one would cut it, and the other would choose which piece to take. In theory, this encouraged the brother doing the cutting to act fairly. In practice, arguments still ensued. Inspired by these interactions, Güth created the Ultimatum Game. Arguments would ensue from it, too.

Here's how the game works. You're going to be playing with another person who is in the room next door. The other person, you are told, has been given an amount of money, say ten dollars. They have been asked to share some of it with you. If you choose to accept what they offer, you keep that money. The other player will keep the rest. However, you also have the choice of rejecting their offer. If you do, then both you and the other player get nothing. You only play the game once. The offer the other player makes is hence an ultimatum: take it, or get nothing at all.

Let's say you were playing the game, received your offer, and found that the other player was proposing to give you two dollars out of the ten. This means they would keep eight dollars for themselves. Would you take it? This is real money, remember, not Monopoly money. What's going through your head? Conceivably it's something like: *Well, they didn't have to offer me anything at all, so it's pretty kind of them to offer me two dollars. I'll take it.* Or maybe it's: *This is unfair, but two dollars is better than nothing, so I'll take it.* Alternatively, it could be: *This is unfair. I'm going to reject it, to teach them a lesson.* Or perhaps you're thinking: *Ah, so you think you're worth more than me, do you? Well, you can lose your eight dollars, buddy. It only costs me two.*

What did the experts think people would do? Economists and mathematicians expected people to accept any offer, as long as it was more than nothing. Economists viewed the player as a *Homo economicus*, who would act to maximize their payoff. This creature, driven by material self-interest, should take free money when offered it, not reject it. After all, something is better than nothing.

But this is not what Güth and his colleagues found. People frequently acted in ways that did not maximize their material self-interest. They often turned down low offers of free money. Today it is established that around half of people reject offers of two dollars or less out of a ten-dollar pot.[9] By doing so, people (and maybe you too) show that they prefer no one getting anything to themselves getting a little and another getting a lot.

When Güth first reported his findings, economists tended to have one of two reactions. The first was to ask, "Are those students in Cologne stupid?" The second was to wonder if Güth had done the study wrong. Surely people wouldn't turn down free money if they understood what was going on. Some players who made low offers, only to have them spitefully rejected, were also puzzled. One, upset that the other player rejected his low offer, bemoaned, "I did not earn any money because all the other players are stupid! How can you reject a positive amount of money and prefer to get zero? They just did not understand the game!"[10] Despite these reactions, researchers around the world have found similar results to Güth's.

Indeed, a couple of years later, and six thousand kilometers west, another research group independently came up with the idea for an Ultimatum Game. This group included Daniel Kahneman, who would go on to win the Nobel Prize in Economic Sciences. They found the same pattern of results as Güth. What struck them was how the rejection of low offers clashed with economic theory. "It's the resentment, the willingness to punish at cost, that is the whole thing," Kahneman remarked. "The thing that's truly bewildering," he added, was that the idea of *Homo economicus* could "stand for hundreds of years, unchallenged, until someone says 'look at the emperor, no clothes.' The counterexample was trivial."[11]

To be fair to the economists, they were spot-on about the pattern of behavior that players would demonstrate during the Ultimatum Game. They were just wrong about the species in which they would see it. Whatever the reality of the Baader-Meinhof prison deaths, we can at least conceive of a spiteful, suicide-based explanation. Yet such a justification would be baffling to our closest genetic relatives, chimpanzees. Admittedly, chimps generally lack interest in 1970s West German politics, and their grasp of the English language can most charitably be described as tenuous. However, a more fundamental barrier to their understanding of the potentially spiteful nature of the Baader-Meinhof suicides is that chimps do not act spitefully.

A chimp will harm another chimp that has hurt it, but it will not pay to harm a chimp who has divided up spoils unequally.[12] Chimps will accept any nonzero offer. Our other close relative, the bonobo, does the same.[13] In contrast, we humans are likely, both literally and metaphorically, to fling the banana back.[14]

Admittedly, economists' predictions of how people would behave in the Ultimatum Game did approximate the behavior of a specific group of Westerners. Members of this group are more likely than the average person to accept low offers in the Ultimatum Game.[15] Yet these are strange people. They seem to struggle with the very concept of fairness itself. Researchers asked members of this tribe what a fair financial contribution to an economic game was. More than a third either refused to answer the question or gave a complex, uninterpretable response.[16] They were also half as likely as regular people to say they were concerned with fairness when making their decisions. The researchers who studied this odd group concluded that "the meaning of 'fairness' in this context was somewhat alien for this group." Members of the group are also less likely to contribute to charity. The strange ideas they are exposed to in their long, expensive initiations may explain why.[17] I'm sure you've worked out who I'm talking about by now. Yes, economists.[18]

What can we take from all this, beyond 1970s economists being good with chimps but bad with people? Before we accept the results of the Ultimatum Game, we need to ponder whether the results are what they appear to be. They seem to show that large numbers of people will act spitefully. But perhaps people only turn down low offers because the amount of money they forgo is trivial to them. Someone might turn down an offer of 10 percent of a $10 pot. After all, what can you buy with a dollar? But would they turn down 10 percent of a $100 pot? Ten bucks is dinner.

To answer this, the American economist Elizabeth Hoffman and her colleagues blew $5,000 running the Ultimatum

Game. In each game, students would divide $100 of real money.[19] There was a lot of pizza on the line. As usual, each student played the game only once. Imagine you are being made an offer. Would you accept $10? What about $30?

Hoffman and colleagues found that some students still acted spitefully even if it involved turning down double-digit offers of free money. Of those offered $10 out of the $100, 75 percent turned it down. Even when students were offered $30, nearly half still turned it down. One student who made an offer of $30 scribbled on the offer paper that the other person shouldn't be a martyr as it would be the easiest money they had ever made. This did not have the intended effect. The other player rejected it and passed back their own note. "Greed is driving this country to hell," it said, "become a part of it and pay."[20]

You might argue that $30 is still not a lot of money. Lucky you. In which case, we can look at what happened when the Ultimatum Game was played with students in Indonesia.[21] Here the pot to be split was equivalent to three times what the average student spent in a month. Think about how much you spend per month. Would you turn down free cash that amounted to even 10 percent of it? Some Indonesian players did. Around one in ten turned down offers of 10 to 20 percent of the pot, thereby giving up significant amounts of free money.

Before concluding that this tells us something important about people, let's consider another objection. Perhaps spiteful behavior in the Ultimatum Game is limited to people from what have been called WEIRD (Western, educated, industrialized, rich, and democratic) societies.[22] A remarkable series of studies, led by the anthropologist Joseph Henrich, has found that people in different societies act very differently when playing the Ultimatum Game. This work began in 1995 when Henrich traveled to Peru to run the Ultimatum Game with the Machiguenga, indigenous people who live in the Amazon basin, not too far from Machu Picchu.[23] It quickly became apparent to Henrich that the Machiguenga played the game very differently from anyone who had played it before.

There were hardly any rejections of low offers. Only one out of the ten Machiguenga people who were offered a split of 20 percent or less of the pot turned it down.[24] "It just seemed ridiculous to the Machiguenga that you would reject an offer of free money," recalled Henrich. "They just didn't understand why anyone would sacrifice money to punish someone who had the good luck of getting to play the other role in the game."[25] They would have made fine economists.

Henrich went on to run the Ultimatum Game in fourteen other small societies, including in Kenya, Ecuador, Paraguay, and Mongolia, publishing his results in 2001. He reported extreme variations in how people responded to low offers across those societies.[26] When the Aché, a nomadic South American people, played it, no one rejected offers of two dollars or below. Yet 80 percent of the Hadza people of Tanzania, who live in the Central Rift valley, rejected offers of two dollars or below. Although the extent of spitefulness may vary between cultures, for reasons we will come to later, it is not uniquely Western.

The results of the Ultimatum Game seem pretty irresistible: spiteful behavior is common. Indeed, spite seems itching to get out of us. And when our inhibitions are lowered, spite practically pours out. One way to see this is to examine how drunk people play the game. This was at least one of the reasons why, in 2014, researchers in Pittsburgh could be found hanging around outside bars until three a.m.[27] They learned two things. The first was that the question "Do you want to come to my mobile laboratory to play the Ultimatum Game?" isn't as creepy as it sounds. Two hundred and sixty-eight people agreed. Of these, seventy-seven were three sheets to the wind. They had a breath alcohol content of .08 or more, making them legally intoxicated. The second thing the researchers learned was that drunk people rejected more low offers in the Ultimatum Game than sober people. Taking the brake off our brains releases the spite within.

Another group of researchers investigated a similar question. However, they opted for an approach that gave them less chance of being vomited on.[28] Their study started from the observation that self-control is like a muscle.[29] If we use it intensely for a while, it weakens. The researchers asked two groups of people to play the Ultimatum Game. First, though, they fatigued the muscles of self-control in the members of one group. They did this using what is called a Stroop task: you are shown the name of a color (e.g., "yellow"), but the ink used to write the word is another color (e.g., blue). Your task is to say the color of the ink as quickly as you can. Now, if you see the word "yellow" staring you in the face, you automatically want to say "yellow." But if the word is written in blue ink, you have to wrestle with yourself to make "blue" come out of your mouth instead. Under time pressure, this is a wearing task. The group that just walked up and played the Ultimatum Game rejected 44 percent of unfair offers. However, the group whose self-control had been worn down by the Stroop task rejected 64 percent of offers. Sapping self-control makes us more spiteful. Spite is automatic and effortless.

Admittedly, such findings come from an artificial laboratory game. What about spite in the real world? When we look around, do we see anything other than minor acts of personal spite? How far can spite go? Do we see spite in the business world, where maximizing shareholder value is required? Does spite manifest on the political scene? When push comes to shove, are people really prepared to be spiteful, or does self-interest win out? Let us go in search of spite. We will start looking where evolution suggests conflict should be most evident: in matters of sex and reproduction.[30]

———

IN 2013 ALAN MARKOVITZ BOUGHT a beautiful lakeside house next door to his ex-wife. Markovitz's backstory is best summarized as colorful. Author of *Topless Prophet: The True Story of America's Most Successful Gentleman's Club Entrepreneur*, he had survived two shootings, one by a stripper, another by a cop.[31] Markovitz believed that his ex-wife was living with a man she had been having an affair with during their marriage. As a result, he spent $7,000 on a twelve-foot-high bronze statue that he placed on his back porch. He shelled out another $300 on a spotlight, to make sure it was visible at all times. The statue, which he positioned facing his ex-wife's house, was of a hand with a single raised finger. You can guess which one.[32]

Millennia ago, the Greek philosopher Aristotle defined spite (*epêreasmos*) as hindering the wishes of others, not so that one may obtain something for oneself but so the other may *not* obtain something.[33] Had he not been a philosopher, Aristotle would have been an insightful divorce lawyer. His definition of spite accords with what we see in many acrimonious separations and divorces.[34]

Dying relationships frequently emit spite. Spouses may destroy their assets to keep them out of their partner's hands, even though doing so makes them worse off too. Even the perception that a partner could be acting out of spite can be damaging. Some abused women are afraid to speak out for fear their partners will portray them as motivated by spite.[35]

Spousal spite can be fatal. Cordula Hahn and her husband, Dr. Nicholas Bartha, lived in a four-story nineteenth-century townhouse on the Upper East Side of New York.[36] For Bartha, it was more than a house. He was a Romanian immigrant, and the home symbolized his achievement of the American Dream. When they divorced, Bartha did not want his ex-wife, whom he described as a gold digger, to get her hands on his dream. A court viewed things differently. It ordered Bartha to sell the house to pay his wife over $4 million. On July 10, 2006, Bartha rigged up a hose to the building's gas supply. He unrolled it down into the townhouse's basement, which filled with gas. The house, with "Dr. Boom" (as the *New York Post* dubbed him) in it, exploded. Bricks rained down on one of New York's wealthiest streets. Bartha was pulled alive from the ruins. After six days in a coma, he died. Spite cost him everything.

Property is not the only collateral damage caused by spousal spite. Children can get caught in their parents' spite when they are used as weapons in vindictive spousal battles.[37] A parent may seek custody of a child simply to spite their partner.[38] Unsurprisingly, this can damage the child's development.[39]

Spousal spite can even lead to infanticide. In *The Selfish Gene*, the biologist Richard Dawkins mentions spite in the context of a hypothetical woman abandoned by her husband. "It is no comfort to her," writes Dawkins, "that the child contains half the male's genes too, and that she could spite him by abandoning it. There is no point in spite for its own sake."[40] Yet some parents clearly find a point in such actions because Medea is not just a myth. Nor is she always female.

June 20, 1999, was Father's Day in the United States. As dawn broke, police in Franklin, Indiana, responded to a call at the house of husband and wife Ronald and Amy Shanabarger.[41] The previous night Amy had come home late from work. She'd gone straight to bed, assuming that their baby son, Tyler, was asleep. He was growing up well. Recently he had started to sit up on his own, and he loved playing peek-a-boo with a washcloth. "How's Tyler?" Amy had asked Ron before falling asleep. "Fine," Ron had whispered.

When the day dawned, Amy went into Tyler's room. She found him lifeless, facedown in his crib. Tyler's body was transferred to the Indiana University Medical Center for autopsy. The pathologist determined that his death was consistent with sudden infant death syndrome (SIDS), a diagnosis that can be given if a child less than a year old suddenly and unexpectedly dies. Only later did the truth come out.

The events that set Tyler's birth and death in motion began three years earlier. Ronald and Amy were not yet married. Amy was on a cruise with her parents when Ronald's father died. Ronald asked Amy to return immediately to attend his father's funeral. She said no. Ronald was not happy. He told his coworkers that he didn't know if he could forgive her. Nevertheless, the following year he married Amy. On Thanksgiving Day 1998, Tyler was born. Ronald did not visit his wife and newborn son in the hospital. Something was not quite right. Those who came to Amy and Ronald's house on that fateful Father's Day in 1999 also noticed this "something." Amy was sobbing when the police arrived, but Ronald was cold, distant, and unconsoling. When Amy's parents came to support their daughter, Ronald gave her father a Father's Day present: a knife.

What was wrong emerged two days after Tyler's death. After his son's funeral, Ronald took his wife to one side. Tyler had not died naturally, he told Amy. Ronald revealed that when Amy had been at work, he had wrapped Tyler's head in cellophane. As Tyler suffocated, Ron got something to eat, brushed his teeth, and wandered back into the room twenty minutes later. He removed the wrap, placed Tyler facedown in his crib, and went to bed. Ronald said he did this because Amy had refused to cancel her holiday to attend his father's funeral. Ronald had married her, gotten his wife pregnant, then waited until the first Father's Day his new son saw before killing him. A judge sentenced him to forty-nine years in prison. This story is not unique. There are numerous examples of parents killing their children to spite their part-ners.[42] We will not pursue the details.

People also kill themselves to spite others. "Spiteful sui-cides" are a thing.[43] Joshua Ravindran, a university student in Australia, had a very close relationship with his father, with whom he shared a house. Others would later refer to their relationship as "special" and "unique." Sometimes, these are not good ways to have your relationship described. When Joshua told his father that he wanted to move out of the house, a heated argument ensued. The next day Joshua found his father hanging by the neck from a rope. A trial concluded that his father was "manipulative and controlling" and that he could have hanged himself in a "spiteful act" simply to make his son feel guilty about their fight.[44]

Crisis negotiators are intimately familiar with spiteful self-harm. One negotiator recalls how a man threatening suicide with a knife demanded that his wife be brought to the scene. The authorities persuaded her to come. Upon see-ing her, the man said, "Look what you've done to me," and proceeded to cut open his stomach. In another example, a police officer went to his ex-girlfriend's apartment, kicked down the door, and confronted her and her new boyfriend. The officer also shouted, "Look what you've done to me." He did not cut open his stomach, though. He shot himself in the head.[45] One can never safely guarantee that love will not give birth to death.[46]

The spiteful behavior Güth found in the Ultimatum Game is very much present in human relationships. Maybe some of the examples above are best understood as incom-prehensible acts of madness. Indeed, there is no one more likely to succeed with an insanity defense than a mother who has killed her children.[47] Furthermore, specific cases will need to be understood in specific ways. For example, to explain spiteful infanticide we need to use the lens of the evolution of human sexual behavior.[48] Nevertheless, we can still look for general principles for why people are prepared to act spitefully. Spite on the grand scale can then be understood as an exacerbation of mundane spitefulness. If we understand the acorn, we will understand the oak.

WE MIGHT THINK THAT THE business world would put the passions of domestic affairs to one side. Maybe here people do behave like *Homo economicus*, always acting in their own material self-interest. Alas not. Spiteful acts between business rivals turn out to be rather common.

The year 1958 was a good one for selling tractors in Italy. As a result, Ferruccio, the head of an Italian tractor company, not only was able to buy a Ferrari for himself but could afford one for his wife too. However, Ferruccio was by all accounts not a great driver. After he had burned out the clutch on his Ferrari for the fourth time, he decided that he would get his head tractor mechanic to repair it, rather than trek out to the nearby Ferrari factory. The mechanic was surprised to find that the clutch the Ferrari used was identical to the ones Ferruccio's company used on their small tractors.[49] He told his boss. Ferruccio was not happy. He had been paying a hundred times more for the replacement clutch from Ferrari than it cost him for an identical tractor clutch. Ferruccio confronted Enzo Ferrari, the founder of the company that bore his name. The conversation became heated. "You build your beautiful cars with my tractor parts!" shouted Ferruccio. Enzo Ferrari's response poured petrol on the fire: "You are a tractor driver, you are a farmer. You shouldn't complain driving my cars because they're the best cars in the world." It was then that Ferruccio decided to make his own sports car, to show Ferrari how it should be done. This was a risky venture, as his wife told him, repeatedly. But Ferruccio was prepared to take the risk. He would act out of spite. Ultimately, he and the company that bore his family's name were rather successful. Ferruccio's surname was Lamborghini.

Such spiteful acts continue today, even among business titans. Take Warren Buffett, the most famous investor in the world and one of the richest people on the planet.[50] He first invested in Berkshire Hathaway in 1962, when it was a failing textile company. Its management was selling textile mills and using the proceeds to buy back shares. Buffett thought he could make a profit by buying shares, waiting for the company to sell another mill, then selling his shares back to it. Its management asked him at what price he would sell his stock. Buffett promised to sell at $11.50. A few weeks later, the management sent out a letter offering to buy stock. But not at $11.50. They offered $11.38. This difference of one-eighth of a dollar was to prove an expensive mistake. It violated Buffett's moral code. It made him mad. Buffett bought more stock, gained control of the company, and fired the top manager. Yet by having his money tied up in textile assets, Buffett was unable to invest it more profitably elsewhere. He estimates that over the long run his spiteful move cost him $200 billion (yes, billion). As a result, Buffett now advises, "Whether you cut off your nose to spite your face or whatever, if you get in a lousy business, get out of it."[51]

Some successful businesspeople can resist the whisperings of spite. Let's take another example involving Buffett. He coined the metaphor of a moat to describe how big companies had competitive advantages over start-ups. Elon Musk, another of this planet's wealthiest people, and soon to be Mars's wealthiest, did not agree. He felt that new technology could make moats redundant. And he said so, calling the concept "lame." To which Buffett replied, "Elon may turn things upside down in some areas," but "I don't think he'd want to take us on in candy." Buffett owned See's Candies. Musk took to Twitter in response. "I'm starting a candy company & it's going to be amazing," he tweeted. "I am super super serious." One can get many things from Musk today, including cars, rockets, flamethrowers, and rap tracks about gorillas,[52] but candy is not among them.

Spite may seem like a personal matter, but it can also rear its head in negotiations between groups. This includes bargaining between companies and unions. Here, if either side acts spitefully, jobs and livelihoods can be lost forever.[53] In negotiations, understanding what might generate a spiteful response is critical.

It is clear that spite, both personally and professionally, can cost people dearly. Yet it also has the potential to affect events on the world stage. Spiteful people can change all our worlds, not just their own.

SPITE THRIVES IN THE DARKNESS of the voting booth. In the nineteenth century, the French psychologist Gustave Le Bon saw the potential for spiteful voting. Le Bon claimed that spite was involved when the person on the street elected one of their peers to office. He argued they did it "to spite an eminent man, or an influential employer of labour on whom the elector is in daily dependence and whose master he has the illusion he becomes in this way for a moment."[54]

Le Bon's spiteful voting has not gone away. Standard political theory suggests that people vote to increase the chance that their preferred candidate wins. Yet a 2014 study found that 14 percent of people who were indifferent to the available candidates would nevertheless drag themselves to the polls to spite a candidate. If these voters discovered that a candidate had broken promises, they would make the effort to spite them by voting for their opponent.[55]

The pithy term "spite-voter" was introduced in 2004 in an article in a New York newspaper.[56] The article's author, Mark Ames, argued that there was a truth that the left struggled to understand: "Millions of Americans, particularly white males, don't vote for what's in their so-called best interests." Ames's explanation was that these voters were casting their ballots to make those who were happier, prettier, and richer than themselves worse off. They were voting, as Ames put it, "out of spite."

The motivation of the spite-voter, in Ames's view, was not rational self-interest but the defiance of those who were doing better, who knew better, and who had the gall to want to help others lower down the ladder. Ames argued that if there was one "perfect spite-ist president," it was Richard

Nixon. This was a man who, as Ames put it, "looked mean, spoke mean and stomped on the hippies who were having too many orgasms." Writing at the time of the 2004 presidential campaign, in which George W. Bush faced off against John Kerry, Ames had a prescription for how the left could neutralize the spite-voter. All they had to do was "not stir up the wrong bile." Twelve years later, Hillary Clinton would famously refer to half of Donald Trump supporters as "a basket of deplorables." Not only did she stir up bile, but she did so using a blender with no lid.

Spite not only impacts whom we elect. It also affects the policies our elected officials pursue and how we feel about them. Consider policies relating to income distribution. We might expect that as economic times get tougher people would be more in favor of their government reducing the difference between the rich and the poor. Yet the recession of 2008 saw people becoming less keen on governments redistributing income.[57] This seems to be because some people don't want tax cuts that make them slightly better off if the cuts also cause those lower on the social ladder to edge closer to them. We display "last-place aversion."

Just as our aversion to being in last place can make us act spitefully, so can our aversion to others being in first place. Imagine you've been made your country's finance minister for the day. A civil servant enters your office and presents you with two choices. Option 1 is to have the wealthiest people pay 50 percent extra in tax, the revenue from which will allow you to give $500 to every person living in poverty. Option 2 is to have the wealthiest pay only 10 percent extra in tax. Yet this low tax rate will incentivize the wealthy to work more, yielding much more tax revenue and allowing you to give a full $1,000 to every person in poverty. Which option do you choose?

In 2017, the psychologist Daniel Sznycer and his colleagues gave this dilemma to people from the United Kingdom, the United States, and India.[58] Around 85 percent picked the win-win situation represented by option 2. They opted to minimize the pain for the wealthy and maximize the help for the poor. Yet 15 percent of people went for option 1, thereby displaying a decidedly different preference. They maximized the harm to the wealthy and minimized the help for the poor. They made everyone lose. They chose the spiteful option.

———

EVEN IF YOU DON'T CARE for relationships, work, or politics, you should still be afraid of spite. This is because it could well destroy us all. Spite poses an existential threat to humanity. Nick Bostrom, a philosopher at Oxford University, asks us to imagine that humankind's ability to come up with new ideas is akin to pulling balls out of a hat.[59] Most of the balls we have pulled out so far have been white. These are ideas that have benefited the world. We've also pulled out grey balls. These represent mixed blessings. Nuclear fission is a grey ball. It has led to new energy sources but also to dangerous weapons. Bostrom cautions that there is another type of ball lurking in the hat, one not seen to date. This is a ball that, if pulled out, would yield a new invention that would devastate humanity: the black ball.

That we have not yet pulled out a black ball is pure luck. What would have happened, Bostrom asks, if nuclear weapons had turned out to be really easy to make? Imagine the consequences if anyone could assemble a nuke in their shed.[60] Bostrom worries that if a black ball appeared then some people would act in a way that would destroy civilization, even at a high cost to themselves. He calls them the "apocalyptic residual." The apocalyptic residual will be a spiteful person. All that stops a spiteful species from self-destruction is the means. We need to understand and control spite before the first black ball emerges. It is not clear how long we have.

＝＝＝

ALTHOUGH SPITE CONCENTRATES IN CERTAIN people, we know little about who these people are. Preliminary research has found that, on average, men are more spiteful than women, and the young are more spiteful than the old. Those who are spiteful are more likely to be aggressive, callous, manipulative, and exploitative. They are also more likely to be lower in empathy, self-esteem, conscientiousness, and agreeableness.[61]

Spite is associated with the possession of the dark triad of personality traits: psychopathy, narcissism, and Machiavellianism. These traits are limbs of a larger tree called the "dark factor of personality," or D-factor for short.[62] It is the tendency to grab what you value (e.g., pleasure, power, money, and status) while disregarding, accepting, or even enjoying any harm your behavior may cause others. People with high D-factors tell themselves stories to help justify their actions. For example, they tell themselves that they are superior, that dominance is natural and desirable, and that everyone thinks about themselves first, so it is fine for them to do so too.

All this makes spite seem worthy of the contempt we like to heap on it. But there's more to the story. Spite has legitimate benefits. My discussion of the Ultimatum Game focused on what people do when they *receive* low offers. In his 2010 book, *The Rational Optimist*, the science writer Matt Ridley focuses on how people behave when they are the one *making* the offer.[63] In that scenario, people typically make relatively fair offers of just over 40 percent of the pot.[64] "Generosity," notes Ridley, "seems to come naturally." Yet part of the reason people act this way is because they worry the other player has a spiteful side. An ungenerous offer could trigger spite and leave them with nothing. Generosity with a gun to your head is self-interest. Spite may not merely be a response to unfairness. It may be what helps people to be fair at all.

A gun in the right hands can save a life, but in the wrong hands it can end one. Another misunderstanding of spite is to see it as fundamentally problematic. Clearly, it can be wielded for ill. Yet it can also be deployed for justice and as an aid to creativity, as we will see. It is more likely to be problematic in the hands of a specific type of person. When we look closer at the people who spitefully reject low offers on the Ultimatum Game, it emerges that this group comprises two different types of people. One uses spite for egalitarian ends, the other for domination. To understand this, we need to address a fundamental and contentious question about human beings. What type of creature are we?

Two
COUNTERDOMINANT SPITE

A rguments about whether we are an equality-loving, egalitarian creature or a power-seeking, domineering one have raged for centuries. All answers are controversial. This is because, rightly or wrongly, answers have political implications. Someone who believes we evolved to live in dominance hierarchies would see communistic societies as doomed to bloody failure. My view, drawing on the work of the anthropologist Christopher Boehm, is that we have evolved both egalitarian and dominance-seeking tendencies. We are two creatures in one. Which aspect predominates depends on the state of the world. Unfortunately, we forget this. We continually mistake a part for the whole.

We are descendants of the victorious. They echo within us. They prompt us to answer the world the way they did, even if their world is long gone. To know ourselves, we must uncover which of our ancestors' strategies were most successful in dealing with the world they faced. One way to answer this is to dig. However, although the dead tell no lies, neither are they terribly forthcoming with the truth. Thankfully, we have recourse to the living. Many peoples today live in conditions similar to those that our ancestors evolved to deal with. These are hunter-gatherer tribes. They give us a window into our evolutionary past. Not all time travelers need a DeLorean.

Contemporary hunter-gatherers face challenges comparable to those that anatomically modern humans faced when they appeared in the late Pleistocene, about fifty thousand years ago. Today, these tribes live in independent mobile bands of twenty to thirty people who are not simply all extended family.[1] They hunt and share the meat of the animals they kill. Christopher Boehm has examined over three hundred such societies. His conclusion about their structure is clear: today's hunter-gatherer societies are egalitarian. From this Boehm infers that in the late Pleistocene, "very likely almost all the humans on this planet were practicing such egalitarianism."[2]

This egalitarianism manifests as a failure to tolerate group members who seek power. The group will not put up with people trying to dominate and bully.[3] As one of the African !Kung hunter-gatherer people puts it, "When a young man kills much meat, he comes to think of himself as a chief or a big man, and he thinks of the rest of us as his servants or inferiors." Such behavior is troubling to the !Kung. "We can't accept this. We refuse one who boasts, for some day his pride will make him kill somebody."[4]

An important caveat is added by the anthropologist Richard Wrangham. As he notes, in small-scale hunter-gatherer societies, "Egalitarianism is primarily a description of relationships among men, particularly married men."[5] Men may behave in an egalitarian way with each other, but not necessarily with women. As Wrangham points out, equality among the Ju/'hoansi hunter-gatherers is said to apply to all adults, but the punishment for a man who beats a woman may be minimal. Similarly, Wrangham observes that Tanzania's Hadza hunter-gatherers are described as egalitarian, yet if there is only limited shade in an area, the men get it while the women sit in the sun. Wrangham gives many more examples of men's appalling treatment of women in these "egalitarian" societies.

Counterdominant behaviors maintain this structure. Hunter-gatherers have ways to crush those who seek to dominate. As Boehm puts it, there is "active and potentially quite violent policing of alpha-male social predators."[6] Hunter-gatherers will even kill group members who try to bully and dominate the group. A study of a Canadian Inuit group describes how one particular menace to the community was "the aggressive, strong man who often took what he liked (including women) by force."[7] Such men did not prosper in the long run. They were often killed, sometimes even at the hands of their own families. *Sic semper tyrannis.*

Egalitarianism hence appears to be "an ancient, evolved human pattern." Possessing a counterdominant attitude of "no one is going to get away with more than I" ensures equal shares of resources such as food.[8] It thereby confers an evolutionary advantage on individuals. Just as our dominant behaviors stretch back deep into the animal kingdom, so do our counterdominant behaviors. They are present, to a lesser degree, in our closest primate relatives. Bonobos will sometimes form coalitions of subordinates who will attack a dominant male to kill or banish him.

However, human counterdominant behaviors differ from those of our primate cousins. When chimps exhibit counterdominant behaviors, one such chimp takes over from the old alpha. Yet humans can be satisfied with justice being done. They don't necessarily take the place of the other. As David Erdal, a former anthropology researcher at the University of St. Andrews, observes, "This makes human counter-dominance something new, qualitatively different in motivation from a simple extension of primate common-ancestor politics."[9] This new form of counterdominance is made possible by other unique abilities that we humans have. In particular, we moralize, we wield weapons, and we have language.

Our moral emotions cause us to react strongly to people who act unfairly and try to dominate others.[10] The American psychologist Jonathan Haidt proposes that we have evolved a specific moral sense relating to the violation of liberty.[11] This, he argues, arose in response to the challenge of living in small groups with others who, given a chance, would try to dominate you. People who failed to experience this moral emotion (and, as we will see, were not prepared to spite others in response) would be less evolutionarily successful. Whether our egalitarianism is driven more by a love of fairness or a hatred of domination is an important question. Haidt opts for the latter. I agree. We will come back to that inquiry later.

Our ability to make tools, specifically weapons, also helps us to act in a counterdominant manner. It is hard for us to kill each other with our bare hands. If we could, there would be no need for Krav Maga. Even our toolless chimpanzee cousins, who are much stronger than we, find it hard to kill each other one on one.[12] Weapons allow weaker humans to take on stronger ones successfully. Our ability to kill is aided by our understanding of the concept of death. Chimps may wound other chimps, leaving them alive. But humans know what death is. And they aim to inflict it.[13]

Despite a potential role for weapons, language seems to have been the crucial factor that allowed counterdominance.[14] Language permitted small groups to coordinate with each other to bring down tall poppies, to test out ideas, and to gossip.

These unique factors that allowed groups of human males to bring down other human males who tried to dominate the group, creating a more egalitarian society, appear to have had a remarkable effect on our species. Richard Wrangham argues that they led us to self-domesticate.[15] Not only have we humans domesticated wolves into dogs; we also seem to have domesticated ourselves, changing an aggressive ape into a calmer, more tolerant one. Wrangham argues in favor of an "execution hypothesis," which builds on an observation made by Charles Darwin that "violent and quarrelsome men often come to a bloody end."

What Wrangham proposes is that males who were more aggressive and tried to dominate the group were executed by groups of other males. As a result, the genes of the less aggressive and more docile males survived into the next generation.[16] In the absence of a tyrannical authority, social norms became the new ruler of society. They were enforced by what has been termed the "tyranny of the cousins" or, as Wrangham puts it, the "tyranny of the previous underdogs."[17] People who did not conform to the rules were punished by the group. This explains our powerful urge to conform and the ingrained desire to act to bring down those who do not.

That we have an egalitarian side may be surprising to some. The idea that humans are set up for a hierarchically organized way of life is almost ubiquitous in popular culture.[18] This view is widespread because it is also correct.

Social animals face the problem of how to live in groups where they have to compete for resources and mates. One common solution is to form dominance hierarchies. We find these in the sea, in crayfish and lobsters. We find them in the trees, in chimps and baboons. We find them in the air, in bats and birds. And we find them on the land, in lions and wolves. Dominance hierarchies are both ancient and pervasive. Humans, whether we be swimming, climbing, flying, or walking, carry this legacy within us too. Put us into a group and within minutes we form a hierarchy.

In dominance hierarchies there is a pecking order. Individuals know where they are in the pecking order and defer to those above them. The benefit of this state of affairs is that potentially injurious conflicts are avoided. Everyone profits.

An individual at the top of the hierarchy is said to be dominant. Given the reproductive benefits of holding this position, we naturally have a dominance-seeking side. In many species, physical prowess gets one to the top of the hierarchy. We can think of deer with giant antlers locked in contest. But when we consider our closer relatives, chimpanzees, achieving dominance is not solely about physical strength. Two weaker males may work together to bring down an alpha. In humans, this tendency is elaborated further. We show both "aggressive dominance" and "social dominance."[19]

People with high levels of aggressive dominance demand their own way, take what they want even when it causes conflict, are aggressive, and put others in their place. They use Machiavellian tactics, including deceit and flattery. In contrast, socially dominant people tend to use reason to persuade others. They are confident, happy talking in front of a group, good conversation starters, and like responsibility; others turn to them for decisions. Socially dominant people learn by copying other successful people, whereas aggressively dominant people tend not to use social information in their decision-making.

Humans have evolved numerous adaptations that allow us to live in dominance hierarchies. We understand the rules of hierarchy from an early age. We know whom we must get permission from and whom we are obliged to. We crave a high position within a hierarchy. Indeed, status seeking is a fundamental human motive.[20] We can see differences in the status of others even before we can talk. Like monkeys, we pay great attention to the status of others.[21] Monkeys will give up a treat of sugary cherry juice just for the chance to glimpse an alpha monkey.[22] If you think we are much different, then just go into the magazine section of any shop.

Recognizing and attending to status is beneficial. It helps the lowly learn the secrets of those on high. *Keeping Up with the Kardashians* may be more pedagogy than pap. We pay more attention to the faces of high-status people, and we remember them better. This helps the powerless seek the protection of the powerful. Placing importance on status happens regardless of people's culture, gender, or age.[23] It is a universal.

We are hence a creature endowed by evolution with both counterdominant and dominance-seeking sides. Erdal expresses this well. We possess, he says, "combinations of contradictory dispositions: to get more and at the same time to stop others from getting more; to dominate, and to stop others from dominating. . . . The conflict is deeply integrated in our psychology."[24]

Given that we have both dominant and counterdominant sides, the obvious next question is: what factors influence which side is in the ascendancy? Boehm argues that these factors include how people feel about hierarchy, the degree of centralized command and control needed in society, and the extent to which subordinates can control those above them. When most humans came to live in settled agricultural societies, around ten thousand years ago, egalitarian hunter-gatherer societies gave way to more hierarchical, domineering societies. This was because, as Erdal observes, the new environment disabled our counterdominant tendencies.[25] Humans now lived in larger groups, had private property, and recognized the legitimacy of chiefs.[26] The availability of large surpluses of storable food allowed people to buy protection and fend off counterdominant resistance.

How does all this relate to spite? My argument is that both our evolved counterdominant and dominance-seeking tendencies can create actions that fit the definition of spite. Our counterdominant side doesn't like our being behind. It encourages us to bring down others, even at a cost to ourselves. It may know it's safest to be quiet, but it can't help telling the loud-mouth bully to shut the fuck up. It wants to pull down the powerful rather than elevate itself. I call this pattern of behavior "counterdominant spite." The counterdominant part of our nature encourages us to support ideas and ideologies that attenuate hierarchies, such as universal human rights, multiculturalism, and diversity.[27] It pulls us to the political left.

Not only does our dominance-seeking side dislike our being behind; it actively prefers us being ahead. It will pay a cost to harm others if doing so leads to a relative gain. It will encourage us to drop a rung down a ladder if it means another falls further. I call this "dominant spite." The dominance-seeking side of our nature encourages us to support ideologies that promote the existence of hierarchies, such as nationalism, the Protestant work ethic, and free-market liberalism, and to hold problematic attitudes that legitimize hierarchies, such as racism and sexism, as well as anti-Semitic and anti-immigrant sentiments.[28] It will pull us politically right.

Let's now dive deeper into these two types of spite.

———

THE ULTIMATUM GAME SHOWED US a group of people who behave spitefully. Recently, researchers have discovered that this group is made up of two very different types of individuals, a fact that was revealed when people who behaved spitefully in the Ultimatum Game were asked to play another game.

The Dictator Game is like the Ultimatum Game. It also involves splitting a pot of money. Yet in the Dictator Game, the proposer (the person who chooses what proportion of the pot to offer the other player) is in a different situation. They are told that the other player can't choose whether to accept or reject their offer. They will just be stuck with it.

As the proposer no longer needs to fear a spiteful response, the only pressure on them comes from their own internal moral compass. It turns out that the moral compasses of people who act spitefully in the Ultimatum Game can point in two radically different directions. When such people play the Dictator Game, some offer fair splits and some do not.[29]

Let's focus first on those cooperative souls who offer fair splits in the Dictator Game. It seems reasonable to suppose that their rejections of low offers in the Ultimatum Game were due to their feeling unfairly treated. We find support for this idea when we listen to people saying why they would reject a low offer in the Ultimatum Game. One offered the following reason: "I don't consider myself a 'spiteful' person at all . . . just fair and expecting the other to be fair."[30]

The economist and former wrestler Ernst Fehr argues that fair-minded people reject low offers in the Ultimatum Game because they are prepared to pay to punish someone who has violated a community standard of fairness.[31] When they do so, argues Fehr, the person they punish will act more fairly in the future. When lots of people do so in lots of different situations, society becomes more cooperative.[32] As a result, spitefully rejecting a low offer isn't an antisocial behavior. It is prosocial. It is the sort of counterdominant behavior that sustains the egalitarianism seen in hunter-gatherer societies. Frankly, these people are heroes. They are John McClane in *Die Hard*. They are Jack Bauer in *24*. They are Batman.

The existence of individuals who behave this way inspired the theory of strong reciprocity.[33] Reciprocity means you repay like with like; you are nice to the nice and nasty to the nasty. Weak and strong forms exist.[34] Weak reciprocity is where people reciprocate only when it benefits them. The weak reciprocator is all about maximizing their gains. They aren't going to punish people if it means they have to pay a cost to do so. Spite is not in their repertoire.

In contrast, strong reciprocators will repay harm with harm, even when it costs them to do so. As the American economist Herbert Gintis explains, a strong reciprocator is predisposed to cooperate with others, and so will begin by acting fairly. However, they will also punish uncooperative others, even if it costs them.[35]

Such people do not behave like a *Homo economicus.* They are not maximizing their immediate material self-interest. Gintis and his colleague Sam Bowles propose that such people can instead be called *Homo reciprocans.*[36] Their spiteful behavior in the Ultimatum Game is called "costly punishment." They pay a personal price to punish another.

Costly punishers are not saints. They don't think, *I must take the burden upon myself to punish this person for the greater good.* If they did, we would expect the saintliest people in our society, such as those who donate organs while still alive, to be particularly likely to reject low offers in the Ultimatum Game. But they aren't. When people who had donated kidneys played the Ultimatum Game, they did not reject more unfair offers than nondonating people.[37]

Being more altruistic doesn't make you more likely to engage in costly punishment, but holding doors open for others does. The more likely you are to conform to norms of niceness, the more likely you are to reject low offers in the Ultimatum Game.[38] A tendency to conform to helpful norms, such as holding doors open for others, is a powerful social force. It encourages people to behave for the good of society in the face of selfish temptations. Conformity helps solve a problem known as the "tragedy of the commons."[39]

If we have one large piece of common land and we all maximize our self-interest by letting our sheep gorge on it, we will destroy it. If, on the other hand, we each restrain ourselves, grazing our animals for only a few hours a day, the land will sustain us all. You don't have to vest power in a central authority to do this. If it becomes normal to graze your sheep for only a few hours each day, most people will feel compelled to conform. Today, we frown on conformity, often justifiably.[40] Yet it is also one of the basic ways that humans reap the benefits of living in societies. When we conform to a norm of punishing the unfair, we all benefit. And the costly punishment of unfairness is a prosocial norm. We think this is how most people act. We also believe it is how people *should* act.[41]

Of course, for people to act this way, they must first perceive unfairness. To understand spite in the form of costly punishment, we need to understand how we judge fairness.

———

SOME THINGS MATTER MORE THAN money. If they didn't, we would accept any offer in the Ultimatum Game. Something else comes into our decision-making. We don't just care about how much we get. We worry about how much everyone else gets too. We are averse to inequity.[42]

Obviously, we dislike receiving less than others. What is less obvious, at least if we view people as purely selfish creatures, is that we can dislike receiving *more* than others. Our brain, which, thanks to evolution, is often wiser than we are, whispers that getting ahead might be a bad thing. It knows this could trigger a counterdominant backlash. It hence tries to deter us from such behavior by making us feel guilty. This partly explains why, when people could give nothing to the other player in the Dictator Game, the majority still give something.[43] Of course, not everyone does. We will come to them in the next chapter.

Whether everyone gets an equal share plays into our judgements as to what is fair. But it doesn't determine those judgements, in adults at least. As we age, we realize that inequality can be fair. We now place more weight on intentions. If you tell a child playing the Ultimatum Game that the proposer had no choice but to offer them two dollars from a ten-dollar pot, it doesn't change their acceptance rates much. Yet knowing this makes adults more likely to accept the two dollars.[44] Adults can move their attention away from the offer being unequal and onto the proposer's intent.

If the proposer's intention doesn't appear malign, we are less likely to spite them. If they accompany their low offer with a note offering a sincere apology, spiteful rejections decrease.[45] Likewise, if they mask their intentions, they are less likely to receive a spiteful response. One way to tweak the Ultimatum Game to allow the proposer to hide their intent is by letting them give the other player a shower of personal information along with their offer. This makes it harder for the other player to decode the proposer's intentions. Confusion is sown. As a result, people are less likely to reject low offers.[46]

Similarly, if the proposer does not have intentions, a low offer is less likely to trigger spite. How would you react if told a computer had randomly generated the lousy two-dollar offer you had received? Clearly, there can be no unfair intent in this situation (unless you've been watching too many movies or TED talks about malign AI systems).[47] Getting a low offer from a computer, rather than a person, radically changes people's behavior. The computer's lack of intent avoids tripping our counterdominant side. Normally, about 70 percent of people will reject low offers. Yet when a computer randomly offers a low amount, we see the complete opposite reaction. Now about 80 percent of people will accept the offer.[48] No intent, no foul, no spite.

Our culture sends powerful signals about what is fair. Different cultures have different norms related to sharing. This leads to substantial cross-cultural differences in how people respond to low offers in the Ultimatum Game. The Hadza people of Tanzania are reluctant to share. They view sharing as tolerated theft. Yet strong social sanctions force them to share. The Hadza are therefore keen to punish unfair offers, leading to high levels of spite in the Ultimatum Game. In contrast, the Aché people of Paraguay freely share food.[49] As a result, they don't need to punish. Their levels of Ultimatum Game spite are low.

Even within a culture, if you jiggle people's expectations of what is fair, you alter how spitefully they behave. Western players enter the Ultimatum Game with an idea of what a typical offer should be. However, if you reduce their expectations of what will be presented, they will be more likely to accept a low offer.[50] If unfairness becomes the new norm, people are more likely to accept it.[51] The best way to get away with murder is to normalize it. That isn't a tip.

Perceptions of deservedness are another essential element of fairness judgements. When goods are shared, we consider whether those who got more deserved it. If we think they didn't, we can be inclined to act spitefully. "Money-burning" experiments show this in action.[52]

In these studies, you play an anonymous betting game against other players. You can see how everyone is doing and how much money they have. You soon notice that the computer is letting some other players make larger bets than you can. They seem to have an unfair advantage. By the end of the game, everyone has different amounts of money. Then you see other players being given free money by the computer. They haven't done anything to deserve this. Unfair!

You now get a choice. You can take the money you have won and walk away. Alternatively, you can give up some of your winnings to reduce the winnings of other players. In the language of the experiment, you can pay to burn their money. What would you do? Remember, the game is anonymous, so you don't need to fear retaliation. It turns out that a whopping two in three people pay to see other people's undeserved gains destroyed. Undeserved gains can trigger a spiteful response from others.

This willingness to pay to destroy the assets of others is very human. When the psychologist Keith Jensen and his colleagues ran the equivalent of money-burning experiments with chimpanzees, they failed to find spite.[53] They put a banana on a table outside chimp #1's cage, out of its reach. If the chimp did nothing, then after fifteen seconds the table would slide over to chimp #2's cage. Chimp #1 had an option of pulling a rope to spitefully break the table and stop chimp #2 from getting an undeserved free banana. Yet Jensen found that chimp #1 was not bothered about the other chimp receiving the banana. Whether the table slid to the cage with chimp #2 in it or to an empty cage didn't affect the rate at which chimp #1 pulled the rope and broke the table. Spitefully destroying others' unearned gains seems distinctly human.

———

We pay to punish unfairness because justice feels good. Really good. Our brains react to the chance to administer justice as if we had been given the chance to take cocaine. It is as though we are addicted to justice. Flip through your local TV listings and see how many shows are about quests for justice in one sense or another. As if to make my point, as I write this, I have the Rambo film *First Blood* on in the background.

The Swiss neuroscientist Dominique de Quervain and his colleagues have shown us how the brain reacts to the anticipation of justice.[54] They used an MRI scanner to look at the brains of people playing a game while they were considering whether to punish another player for acting unfairly. The players' brain activity was again similar to that of drug users anticipating taking cocaine. They were getting a rush from the anticipatory joy of justice.

Yet, as any moviegoer knows, justice comes at a price. Is this a cost we want to pay? De Quervain's study also saw people's brains pricing justice. Two key brain regions lit up when people mused about whether or not to inflict costly punishment. The first was their ventromedial prefrontal cortex, which is involved in juggling different goals and controlling anger. The second was their medial orbitofrontal cortex, which is involved in making choices between different rewards. Choosing whether or not to punish at a price can be tricky. Often your brain helps you out by giving you a hint, which comes in the form of an emotion. The emotion in question is related to the impulse to hurt someone who makes unfair offers. The emotion is anger.[55]

▬▬▬

ANGER IS A CRUCIAL FACTOR that converts a perception of injustice into spiteful behavior.[56] Whether in a slighted spouse, a screwed-over businessperson, or a betrayed voter, injustice lights the tinder of spite by striking the match of anger. If the regions of your brain associated with anger light up when you receive an unfair offer in the Ultimatum Game, you are likely to reject the offer.[57] Anger helps explain many findings related to patterns of spiteful behavior in the Ultimatum Game. For example, we saw in Chapter 1 that the young are more likely to act spitefully than the old. This is probably because the young tend to be angrier than the old.[58]

Our emotional response to unfairness is not limited to anger. Another basic human emotion is also activated. If someone gets an unfair offer in the Ultimatum Game, their facial muscle reaction shows a characteristic pattern of disgust.[59] Regions of their brain associated with disgust light up.[60] The combination of anger and disgust creates moral outrage.[61] Unfairness doesn't just anger or disgust us. It outrages us.

Our outrage at unfairness can be so powerful that we will pay to harm the culprit even when we know that our punishment won't affect them at all. Researchers found this using a variant of the Ultimatum Game. In it, if you reject the proposer's offer, the proposer will keep their money but you will lose yours. This situation renders spite impotent. It seems logical to expect that no one would reject a low offer under these circumstances. The only person you will hurt is yourself. Yet researchers found that 40 percent of people still spitefully rejected two-dollar offers from a ten-dollar pot in this scenario.[62] We will stand in the rain to curse the clouds.

The importance of anger to spite is illustrated in studies showing that reducing anger reduces spite. There are a range of ways to do this. One way is chemical. Benzodiazepines, such as Valium and Xanax, reduce the activity in people's neural anger centers: their amygdalae. Giving these drugs to individuals who are playing the Ultimatum Game reduces the activity of their amygdalae when they get low offers and halves the number of times they reject the offers.[63] A more natural way to reduce anger at unfairness is to take a small dose of time. Introducing a ten-minute delay between people receiving a low offer in the Ultimatum Game and their choosing whether or not to accept it leads to a sharp drop in rates of spiteful rejections, from 70 percent to 20 percent.[64]

As one would expect, the more we can control our anger, the more we can control our spiteful behavior. One way to see this is by looking at heart rate variability. The human heart is not constant. Heart rate variability is a measure of the variation in the time gaps between consecutive heartbeats. The higher your heart rate variability is, the better you can control your emotions. As we would expect, those with high heart rate variability make fewer spiteful rejections in the Ultimatum Game.[65]

We can artificially increase our ability to control our anger, which also reduces spite. The neuroscientist Gadi Gilam and colleagues used a technique called "transcranial direct-current stimulation" to induce electrical currents in the brain. Their goal was to increase the activity in a part of people's brains that controls anger (the ventromedial prefrontal cortex) before they played the Ultimatum Game. To make sure people really were getting angry, they created an "anger-infused" version of the game.[66] The researchers gave people low offers along with a provocative note. The note might read, "$2. Come on you loser!" Or, "$2. That's the offer, deal with it." When Gilam and colleagues increased the activity of players' ventromedial prefrontal cortices, players felt that unfair offers were less unfair. Rejection rates of low offers fell from 70 percent to 59 percent.[67]

Whether by culture, chemicals, or current, spite can be controlled.

ALTHOUGH SPITE IS RELATIVELY EFFORTLESS, the brain does a lot of work behind the scenes to make it happen. Indeed, to spite, we have to overcome other elements of our nature: selfishness and empathy.

First, we must be able to turn down a benefit, even if it is small. When receiving a low offer in the Ultimatum Game, we must weigh gaining a small but nonzero amount of money against the cost of tolerating unfairness. To do this, we engage the cost-benefit analysis part of our brain that helps us to control our behavior: the dorsolateral prefrontal cortex.[68] Activity here is necessary to prevent us from acting in our narrow self-interest by taking a couple of bucks. Without it, we would act like chimps and accept minimal amounts of money, oblivious to unfairness.

We typically think of the prefrontal cortex as the rational part of the brain. We view it as conflicting with more ancient, emotional, animalistic parts of our brain. Yet when we receive a low offer in the Ultimatum Game, our rational brain is activated not to overcome our emotions but to let them guide us. Our emotions give us valuable information. Anger may be pushing us to act spitefully because doing so is in our best interests.

The rational part of our brain uses the emotional information provided by anger to set aside selfish motivations and let us turn down free money. We know this thanks to an experiment that used neurostimulation to tie up the dorsolateral prefrontal cortex (remember, that's the cost-benefit analysis part of people's brain) during the Ultimatum Game.[69] In this study, people played the Ultimatum Game with a twenty-dollar pot. The lowest offer players could make was four dollars. In a routine game, only 9 percent of people accepted such offers. But when the cost-benefit analysis region of players' brains was impaired by neurostimulation, they accepted 45 percent of the four-dollar offers. The experimenters checked to see if the brain stimulation had somehow made four-dollar offers seem less unfair. It hadn't. The participants still believed the offer was unfair. They just didn't choose to punish the other person for it. They couldn't see past their immediate material self-interest.

Similar results occurred when neurostimulation was applied to this cost-benefit analysis part of the brain while people were playing a game involving trust. People ran off with the other person's money more often. They were less able to resist short-term temptation. This meant they couldn't build a reputation for cooperation when they played the game multiple times. Having a good reputation would have allowed them to earn more money in repeated play. The people who received neurostimulation were still able to recognize, when asked, that cooperation was in their long-term interest. But what they knew and what they did diverged. They couldn't give up the short-term benefits of cheating to avoid the long-term costs of a loss of cooperation. Such findings explain why reputations are rare in species with less developed prefrontal cortices than ours.[70] They also speak to the importance of reputations to humans.

We may be tempted to conclude that our neural cost-benefit analysis area, the dorsolateral prefrontal cortex, is a general antiselfishness device. We may think it allows us to *resist* selfishly taking short-term gains, helping us act in our longer-term interests. It halts our inner chimp. Yet the reality is slightly more complicated and a good deal more interesting.

Imagine that this area of the brain functioned merely to suppress selfishness. If so, then tying it up using neurostimulation should cause people to act more selfishly in any situation. If such a person were to take on the role of proposer in the Ultimatum Game, we would expect them to make lower, more selfish offers than average to the other player. But this is not the case. It seems that the dorsolateral prefrontal cortex is involved in suppressing self-interest specifically when the person has been treated unfairly and needs to act angrily. It wants to let righteous anger out.[71]

The first hurdle to translating anger into an act of costly punishment is hence overcoming the inherent temptation to get even for a derisory offer. A second, related hurdle is being concerned enough about justice to motivate you to pay to administer it. Being concerned about your own unfair treatment may not be enough to generate spite. You may be too selfish to give up the money on offer. However, if you are concerned about others being treated unfairly, you are more likely to reject a low offer.[72] You are selfless enough to spite.

A final hurdle to translating anger into spite comes from our ability to feel the pain of others. We have empathy. It is harder to punish someone if you feel the pain you are causing. Yet our brains have ways of getting around this. We can see it in the MRI scanner. Imagine playing a game in which some opponents compete fairly against you and others don't.[73] You are shown these opponents receiving electric shocks. Normally, when you see someone in pain, the pain centers of your own brain light up. This is empathy in action. However, when you see unfair players being shocked, the regions of your brain involved in empathy are less active than when you see fair players being shocked.[74] The brain dials down empathy when it sees someone else acting unfairly. It knows that the other needs to be punished and sweeps aside obstacles on the pathway to justice.

Your brain has another, more dangerous trick for getting around empathy. It makes unfair people seem less human. Dehumanization, ceasing to see other people as fellow human beings, is potentially lethal. It helped "colonists to exterminate indigenous peoples as if they were insects and whites to own blacks as if they were property."[75] It turns out that our brain processes the faces of norm-violators in a different way from how it usually handles human faces. This is important because if we don't recognize a face as a face, we fail to get an important cue that what is before us is a fellow human being. When we learn that people have violated a norm, we literally cease to see them as fully human.

Katrina Fincher and Philip Tetlock at the University of Pennsylvania uncovered this phenomenon of "perceptual dehumanization" in 2016.[76] They also found that dehumanization made it easier for people to punish norm-violators. Fincher and Tetlock noted the ease with which their participants could "turn off" their empathy to punish norm-violators. They suggest that this finding may even call into question the idea that empathy is a default part of human nature.

Spiteful people may have lower-than-average levels of empathy to begin with. We know that spiteful people are less able to intuit the feelings, beliefs, and intentions of others.[77] Yet this quality may allow them to enforce rules of fairness more objectively and easily. We may need such people.

In THE FACE OF UNFAIRNESS, our brains not only push us down the road of spite; they clear all traffic in the way. But why do our brains usher us down this costly path? The answer depends on the way we pose the problem. Some researchers think costly punishment yields a societal benefit, by making everyone less likely to act unfairly, but gives no additional benefit to the punisher themselves. This means everyone receives a small benefit from the punisher's actions, but the punisher alone bears the costs. The problem here is that everyone would be better off letting someone else do the punishing.

To make this clearer, imagine you're tenth in a line when another person ambles up and steps into the second spot. Someone needs to put this queue-jumper in their place. But doing so is risky. The queue-jumper could have a knife in their pocket. Maybe you'll just stand quietly, waiting for someone else to intervene. You can then get the benefit of having one fewer person in front of you in the queue— without exposing yourself to any risk.

It thus seems logical to let others do the punishing. Yet as we have seen from the Ultimatum Game, plenty of people are prepared to do the punishing themselves. Indeed, we see this in everyday life. Take a story recounted by the writer Bill Bryson.[78] In 1987, John Fallows went to his bank in London. He had just reached the front of the line when another man, Douglas Bath, ran into the bank. Bath was armed with a gun. He demanded money from the cashier. Fallows was so irritated by Bath's actions that he told him to "bugger off" to the back of the queue and wait his turn. Bath was so surprised that he slunk away; police arrested him shortly afterward. Everyone in the queue benefited from Fallows putting the would-be robber in his place. Yet it was beneficial for them to let Fallows be the one bearing the potential costs of doing so.

We may wonder how such costly punishment evolved. It seems that people with genes that predispose them to perform costly punishment should die out. Others who pay no cost and reap the benefits of your punishing should be more likely to survive in the long run. Eventually, people like John Fallows should get killed and their costly punishing genes die with them.

The situation becomes starker when we consider third-party costly punishment. In the Ultimatum Game, people punish others who have acted unfairly toward them. This is called "second-party costly punishment." But imagine you were watching two other people play the Dictator Game. If you had the chance, would you pay to punish someone who gave an unfair split of the pot to the other player, even though it had nothing to do with you? This would be called "third-party costly punishment." In terms of queue jumping, it would be like yelling at a would-be bank robber who had jumped into the line behind you.[79] Surely no one would do that?

Yet they do. A study led by Ernst Fehr asked people to watch others playing the Dictator Game. At the end, they could pay to punish a dictator who had made an unfair offer to another player. Fehr and colleagues noted that a whopping 60 percent of observers chose to pay to punish dictators whom they saw acting unfairly toward other people.[80]

Such third-party costly punishment seems to be unique to humans. Chimps will harm other chimps who steal their food. However, they will not punish chimps who steal food from other chimps, even if the victim chimp is a relative.[81] In contrast, even young human children will pay to punish third parties who violate norms. The psychologist Katherine McAuliffe and her colleagues found that six-year-old children would give up some of their candy to punish other children who divided candy unfairly with another child.[82] Let me emphasize that again. Our willingness to spite a stranger who acts unfairly toward another stranger is so strong that *children will give up candy to do it*. If this doesn't communicate how powerful our impulse to spite is, nothing will.

However, other researchers have questioned whether the high levels of third-party costly punishment delivered by people watching others playing the Dictator Game may be due to how the experiment was set up.[83] First, by only giving the observer the choice to punish the dictator, there is some implication that they should, which may artificially push people into punishing. Second, the potential punisher knew that other players would be aware of whether or not they dished out punishment. This amounts to an audience watching their actions, which places their reputation as a fair person on the line.

The psychologist Eric Pedersen and his colleagues tried to address these problems. They removed the implicit suggestion that people should deliver third-party punishment, made the punishment anonymous, and reran the study. They found some important differences.[84] First, they found that next to nobody paid to punish a stranger who had acted unfairly toward another stranger in the Dictator Game. Second, the few who did pay to punish did not appear to be driven by anger at a moral violation. It was envy that drove their behavior. They weren't punishing the dictator because they were angry at the unfair acts. They were punishing because they were distressed that the dictator had gained an advantage. Finally, they found differences between how people predicted they would behave and how they actually behaved. People predicted in advance that witnessing unfair behavior by the dictator would make them angry and make them pay to punish. Yet, as we have seen, when presented with that scenario, they did not get angry and did not pay to punish. This seems to represent a split between how we feel we should behave and how we actually behave, something we will return to later in the book.

It hence appears that we pay to punish strangers who exploit other strangers when there is something in it for us. It isn't that the benefits from such actions are equally shared between all in society, with the punisher left with the costs. Instead, the punisher gets their own unique benefits. Spite can function like the colorful markings on an insect. It warns of poison. It signals "here be monsters." As a result, the person you spite improves their cooperation with you and you alone, rather than with everyone. Similarly, other people might see you exacting costly punishment, or be told about your doing so, realize the danger you pose, and increase their cooperation with you specifically.[85] Spite is a signal you are not to be messed with, which can benefit you.

The idea that we gain a direct personal benefit from imposing costly punishment fits with an influential theory of anger. It proposes that the purpose of anger is to get other people to change their valuation of us and therefore act better toward us in the future.[86] In short, if you get pissed at people, they will be forced to care about you more.

The weight that someone places on your welfare, relative to their own, is called a "welfare tradeoff ratio."[87] Others mentally allocate you such a ratio, which they update as new events happen. They use the ratio to guide their decisions about how to behave toward you.[88] Someone might signal that they place a low weight on your welfare, such as by making a low offer to you in the Ultimatum Game. You need to make them adjust the ratio they have given you. By spitefully punishing them for their low offer, you teach them to place more weight on your welfare in the future. In short, if you don't pay to stand up for yourself now, you will get walked all over again later. Spite may be the last weapon of the downtrodden.

Alternatively, spiting a third party may signal that you are a saint rather than a monster. Punishing others at a cost can give you a reputation as a just person. Others now think more highly of you, and benefits, including direct material ones, flow your way. We see this happening when people play a variant of the Dictator Game. Imagine you have seen the dictator offer an unfair amount to the other player, and you pay to punish the dictator. If other people see you do this and have the chance to reward you for it, some will.[89]

In reality, people are unlikely to give you rewards for punishing others directly. Your reward will be that they pick you to work with. This flow of benefits to you doesn't depend on the other person altruistically rewarding you. It only requires them to act in their self-interest to interact with a fair partner.[90] And, our tendency to spite notwithstanding, self-interest is still a strong predictor of how we will act.

It turns out that you only get clear benefits from others when you pay to punish someone who was unfair to someone else (third-party costly punishment). Paying to punish someone who is unfair to you (second-party costly punishment) actually seems to cause problems for you. Such behavior doesn't promote cooperation; it escalates conflicts. As a recent study of punishment and cooperation succinctly concluded, "Winners don't punish. . . . Losers punish and perish."[91]

Indeed, we don't normally esteem people who punish those who have directly harmed them. We see them as vengeful, not fair. As the old saying goes, a judge cannot punish a wrong done to himself. Instead, people who *refrain* from punishing someone who has acted unfairly against them are more likely to be rated as trustworthy and altruistic.[92] There don't seem to be many direct benefits from punishing those who have acted unfairly toward *us*.

In contrast, paying to harm someone who has been unfair to someone else does seem to yield personal benefits. Take people who pay to punish others who have cheated their group, rather than just them individually. In such a case, the punishers are rated as more trustworthy, more committed to the group, and more worthy of greater respect than people who do not undertake such acts.[93] We like those people. They appear to uphold moral norms without self-interest.[94] Heroes stand up for others, not for their own self-interest.

We can understand costly third-party punishment as costly signaling. Peacocks grow huge tails to signal to potential mates that they are so strong they have energy to burn on follicular folly. It shows that they are sturdy birds who will make a fine mate choice. Likewise, someone who punishes another who has harmed someone else sends a costly signal that they are a powerful, just person who might make a good mate. Jumping in to punish someone, when you don't have a dog in the race, is pretty hot.

IF SPITING PEOPLE WHO ARE unfair to you typically escalates problems, why do we see so much of it in the Ultimatum Game? One reason is because the Ultimatum Game does not allow the punished person to retaliate. If laboratory games are changed to let players retaliate, around a quarter of people who suffer punishment will do so.[95] This is a significant disincentive to performing costly punishment.

In the real world, where people can retaliate, directly spiting people who act unfairly toward you is much rarer than we might think from looking at the results of the Ultimatum Game. Anthropological studies of small tribal societies do not show people acting that way to maintain cooperation. Instead, they use alternative forms of counterdominant behaviors that bear a smaller price tag. They use cheap spite.

One way to reduce the personal cost of punishing an unfair person is to dilute the costs, which can be done by spreading them among a group of people. In small societies, groups, not individuals, kill people who have gone too far, thus minimizing individual blowback. Similarly, in the West, people are happier to punish someone if another person is also punishing that person. There is safety in numbers.[96]

Another way to reduce the cost of punishment is to use methods that are cheaper than direct confrontation, including gossip, ridicule, and ostracism. In the real world, people are more likely to use these forms of punishment than direct confrontation,[97] particularly in small societies. Arguments among Inuit can be resolved through their equivalent of rap battles, with each party taking their turn to mock the other in verse.[98] When the social scientist Francesco Guala asked the Hadza people what they did when someone was lazy or stingy, the most common answer was "We move away from them" rather than "We make them leave."[99] The Chaldean community in the United States (Aramaic-speaking

Catholics, originally from Iraq, now living in Detroit) only use the relatively costly punishment of reproaching people for minor infringement of social norms, such as failures to recycle. Serious issues are dealt with by the cheaper punishment of gossiping.[100]

Gossip has two major things going for it.[101] The devil's radio, as George Harrison called it, is effective. Gossip can have a devastating effect on other people's reputations, which makes them behave more cooperatively in the future.[102] Indeed, it seems to be better than direct punishment at promoting cooperation.[103] It is also cheap. The identity of the original gossiper is usually concealed, making it likely they will get away scot-free. Yet gossip is best described as a low-cost—rather than a no-cost—way of punishing another, as it can be costly to your reputation to be perceived as a gossiper.[104]

The rarity in the real world of costly punishment with a substantial price tag suggests that people may only resort to it in the Ultimatum Game because it is the only punishment option they have. In the real world, they could opt for less costly options instead. Indeed, if you tweak the Ultimatum Game so that cheaper punishment options are available, people will often use them. This was found by a 2005 study that gave the responder the additional choice of sending a written message to the proposer.[105] Nearly 90 percent of people chose to send a note when they were offered two dollars or less out of a ten-dollar pot. As you can imagine, most notes were pretty cranky. Being able to send such notes caused spiteful rejections of low offers to fall from 60 percent to 32 percent. One player, who decided to accept the money, wrote, "We should have divided the money equally. Don't be so greedy. People are always out for themselves." If given a choice, many (though not all) are satisfied with a cheaper, written form of punishment.

Written or verbal punishment may be cheaper, but it is also less effective for an individual to wield. In economic lab games, if a single person punishes you verbally, you won't change your behavior. However, if more than one person weighs in verbally, you will become more cooperative.[106] People are reluctant to change their behavior in response to mere disapproval unless they think there is some level of consensus between other people that their behavior is wrong. This is in contrast to financial punishment. Even a single person taking your money can make you cooperate more. Spite is hence like many commodities. The more you pay, the more effective it is.

Nonfinancial punishment is more likely to make people behave better if it comes from a recognized group or institution rather than just an individual. Punishment by an institution also has the benefit of safeguarding societal cooperation by removing the threat of personal feuds and vendettas.[107] But to have this effect, institutions need to be perceived as legitimate.[108] A study in Uganda found that people who were punished by an elected monitor were twice as likely to become more cooperative relative to those who were punished by a monitor who was randomly given the job.[109] If we want to maintain a cooperative society, we must ensure that our institutions are perceived as legitimate.

THERE ARE A COUPLE OF important caveats to the idea that a specific type of person, a *Homo reciprocans*, spitefully rejects low offers in the Ultimatum Game because they are prepared to reciprocate unfairness with costly punishment. The first, which we will discuss more in Chapter 4, is that simply suffering a loss isn't enough to cause people to pay to punish the other party. People respond to losses that mean the other person gets ahead of them.[110] Costly punishment is not merely about reciprocation of harm. It is about wanting to prevent another person from getting ahead of you; it triggers a counterdominance response. This is clearly apparent in the phenomenon of "do-gooder derogation."

If someone is nice to a *Homo reciprocans*, the *Homo reciprocans* should respond in kind. But what if the other person's niceness leads them to gain status? What does the *Homo reciprocans* respond to: the niceness of the other or the status gain of the other? To answer this, imagine you play the following game.

You and three other players each get twenty dollars. You each choose how much to keep for yourself and how much to invest in a group fund. The fund will pay out dividends to all of you. The more money invested in the fund, the bigger the dividends will be. A crucial detail here is that the fund pays out to each of you, *even if you didn't invest any money in the fund*. This means that if you don't invest anything in the fund, but everyone else does, you get paid dividends *and* still have your original money. You will end up with more money than any of the other players. Yet the way to end up with the most money possible is for you and all the other players to contribute as much as you can to the fund.

At the end of the game, players can pay to punish each other for the way they played. For every dollar you pay, three dollars will be taken away from the player of your choice. If you were playing, would you pay to take money away from someone who had selfishly decided not to pay into the group fund? By now we know that many people will undertake such costly punishment. But here's another question, which might seem odd. If you found out that one of the other players had generously invested more than you in the group project, leading you to receive greater dividends, would you pay to punish them? Surely not. The thought probably wouldn't even have entered your head until I mentioned it. You can see where this is going. It turns out Dostoyevsky was right: we are indeed a "creature that has two legs and no sense of gratitude."

In 2008 this experiment was run in sixteen countries by Benedikt Herrmann and colleagues.[111] As expected, when players found out that other people had been selfish and invested less in the group fund than they had, they would often pay to punish them. But what was interesting was what happened when players found out that others had contributed *more* to the group fund than they had: people still paid money to punish them! They punished the generosity of others, even though they had financially benefited from that generosity. This is called "do-gooder derogation." The effect of this punishment of the generous seemed to be counterproductive. It made the generous people contribute less in future iterations of the game. It decreased cooperation. Everyone lost, or at least they appeared to.

The punishment of helpfulness and goodness isn't just an artifact of laboratory games. If we turn to anthropology, we find successful hunters being criticized for catching a big animal that all the tribe can share.[112] In ancient Greece, those whose excellence raised them too far above other citizens could be exiled.[113] In our society, think of vegetarians. Although you may not believe that they are improving the world for you, you can at least recognize that they are choosing to give up meat for moral or altruistic reasons. Yet people who eat meat sometimes see vegetarianism as a public rebuke against them, which can lead them to punish vegetarians.[114] For anyone wanting to make the world a better place, the tendency to punish people who are trying to help others or the world is profoundly worrying.

So why do some people punish the generous? The behavior seems to stem from our counterdominant tendencies being triggered. A less generous contributor may feel that a more generous one has gained a status advantage. Generous acts are associated with status and reputation gains. The generous person is threatening to become dominant. As Voltaire put it, the best is the enemy of the good.[115]

"Biological markets theory," which proposes that people compete to be chosen as a partner to cooperate with, provides a related explanation.[116] One way to become a desirable partner is to be nicer and more generous than other people. However, an alternative strategy is to try to make your competition look bad. Torpedoing a do-gooder may make you, the spiteful person, look relatively better (or at least less bad).

Although such behavior runs the risk of making you appear petty and small, evidence has been found consistent with the idea that we punish the generous to make ourselves appear more attractive as partners.[117] People are more likely to spitefully punish people who have contributed more than they have if they think that others are scrutinizing them as a potential partner for another game.

Do-gooder derogation is a way to bring down the generous. It is a problem for society because it prevents an escalation of generosity.[118] It encourages us to be good, but not too good. We may be adults, but we're still in the schoolyard.

We need to understand what factors encourage the punishment of the generous. Herrmann and his colleagues have provided some answers to this question.[119] They found that people punished generosity more in countries where the rule of law was weaker. People also punished the generous more if societal norms of civil cooperation (i.e., how acceptable people found tax evasion, benefit fraud, or avoiding train fares) were lower. This may be because, in the absence of institutions to perform punishment, the population itself needs to act spitefully to enhance levels of cooperation.

Herrmann and colleagues also found that people punished the generous more in countries that were less equal, probably because getting a leg up in highly unequal societies provides substantial benefits. This is called "reproductive skew." In greatly unequal situations, being at the top of the pile pays handsomely. Our counterdominant side hence wants to pull such people down. Looking around the world today, this does not augur well.

———

To SUM UP WHERE WE'VE arrived, our evolved counterdominant tendencies help us spite people who we perceive are acting unfairly and attempting to dominate us. It makes sense that this costly punishment should be delivered by the sort of people discussed at the beginning of this chapter: fair-minded people who reject low offers in the Ultimatum Game but make fair offers in the Dictator Game. These people abide by the norms they punish others for violating. This is important. If you are punished by someone who themselves complies with norms, you will believe that your norm violation was the reason for your punishment. But if you are punished by someone who does not themselves comply with norms, you will feel they are not enforcing the rules but are trying to get one over on you. You will think they are acting with a competitive aim.[120] You are less likely to accept their punishment. Such a scenario usually motivates the punisher to convince the victim that they are punishing them for moral reasons, not personal gain. The best way to get others to believe your motives is to believe them yourself. The best liar is the one who believes he is telling the truth. As we will see later, we often spite others for a relative personal advantage but fool ourselves into thinking we are doing it for moral reasons.

In the real world, acts of costly punishment are discouraged by the potential for retaliation, leading us to seek cheaper forms of spite. Although a lack of anonymity has historically kept a lid on personal acts of spite, our modern society is very different. We can act anonymously, particularly in the online world, which can create cracks in the dam of accountability that holds back spite. An anonymous social network is destined to become a spiteful social network.

I will return to this issue in this book's conclusion; next, we need to understand the second type of person who acts spitefully in the Ultimatum Game. For a long time, researchers believed that the rejection of low offers in the Ultimatum Game was a result of cooperative people punishing unfairness. Yet this reflected an overoptimistic view of humanity. Spite isn't used merely to promote fairness. As the theologian Reinhold Niebuhr once put it, "Only rarely does nature provide armors of defense which cannot be transmuted into instruments of aggression."[121]

More recent research has found that some spiteful players of the Ultimatum Game will make derisory offers to the helpless other in the Dictator Game. These players aren't just driven by counterdominance; they want to become dominant. They aren't just driven by a hatred of others having more than they; they are driven by a love of having more than others. Their rejection of low offers in the Ultimatum Game can be called dominant spite.

Three
DOMINANT SPITE

"The most imperious of all necessities," claimed the French writer Alexis de Tocqueville, "is that of not sinking in the world." Although we will sacrifice money, we rarely give up status. Imagine that your government proposed to increase the minimum wage. Who do you think would most oppose the idea? It's not those at the top of the earnings scale. Instead, it's those who earn just above the current minimum wage. A substantial number of such people will effectively oppose their own pay increase. The reason is that they don't want those below clambering up, to perch next to them on a new bottom rung. Last-place aversion motivates them to keep their place on the second-lowest rung of the social ladder. They will forgo an absolute gain to maintain a relative advantage. Rejecting the pay raise hence represents an act of dominant spite: harming oneself and another in order to get or keep ahead of them.[1]

As we saw in the last chapter, some spiteful rejections in the Ultimatum Game come from fair-minded people who can be called *Homo reciprocans*. They are strong reciprocators prepared to respond to harm with harm. Yet they are predisposed to cooperate with others. They will offer a reasonable amount to the powerless other player in the Dictator Game. In contrast, another set of spiteful rejections in the Ultimatum Game comes from people who could be called *Homo rivalis*.[2] They are predisposed to act selfishly, not cooperatively. They will make unfair offers in the Dictator Game yet will also act spitefully by rejecting low offers in the Ultimatum Game. They don't do this because they are cooperative people whose counterdominant side has been triggered by unfairness. They do it because by doing so they can gain a relative advantage. They can dominate the other.

For *Homo rivalis*, the Ultimatum Game is not about social exchange; it is about status competition. If they turn down an offer of two dollars from a ten-dollar pot in the Ultimatum Game, they may have lost two dollars, but the proposer has lost eight dollars. They have made a relative gain. A *Homo reciprocans* isn't looking to elevate itself above others. However, a *Homo rivalis* likes getting ahead of others. It would rather rule in hell than serve in heaven.[3]

We can sight *Homo rivalis* in the work of Dutch psychologist Paul van Lange.[4] His research looks at people's "social value orientation," which encompasses the stable preferences we have for how events pan out with regard to ourselves and others. Van Lange's work found that people usually have one of three social value orientations. To see which you are, consider the following game.

You and another player are to be awarded points. The more points you get, the better. You can choose between the following three options:

> Option 1: 480 points for you, 480 points for the other person
> Option 2: 540 points for you, 280 points for the other person
> Option 3: 480 points for you, 80 points for the other person

Which option would you choose?

If you chose option 1, you are "prosocial."[5] This puts you with the majority, as 66 percent of people are of this type. *Homo reciprocans* will form part of this group. If you chose option 2, you are "individualistic." You focus on maximizing your money, like a *Homo economicus*. Around 20 percent of people are of this type. If you chose option 3, you are "competitive." This is what a *Homo rivalis* would do. Only 7 percent of people choose this option.

Option 3 is spiteful. You've paid a cost by not taking option 2, which would have given you the most points. You've also made the other player suffer a cost by giving them the smallest amount possible. You've chosen the option that puts the most daylight between you and the other player. You've maximized not the amount of money in your wallet, but your dominance. The fact that around 7 percent of people act this way fits with the 5 to 10 percent of people who, as we saw in Chapter 1, typically endorse spiteful items on question-naires. This suggests that the questionnaires primarily as-sess dominant spite. Indeed, if we take another look at those questions, we will see that they are not necessarily scenar-ios in which another person has been unfair. They are often scenarios in which the chance to dominate another occurs, albeit at a cost.

Another way to see the difference between counter-dominant and dominant spite is to change the price of pun-ishment in economic games. Most punishment games are designed so that the other player loses three dollars for every dollar you pay to punish. If the other player makes a low of-fer, someone driven by counterdominance will punish them to pull them back to the level of the group. However, a *Homo rivalis* will punish because they make a relative gain on the other by paying one dollar to take away three dollars from them. Theirs is an act of attempted dominance.

Now, if we change the game so that for every dollar you pay, the other person also loses a dollar, then things should change. The person driven by counterdominance should still punish because they want to reduce the payoff of the player who is seeking an advantage. But *Homo rivalis* should no longer choose to punish because doing so doesn't give them a relative gain. Both they and the other player will fall back by a dollar. They can't gain in dominance. They will hence pick up their ball and head home.

When researchers change the price of punishment, this is indeed what we see.[6] In a study by Falk and colleagues, people first played a game to see whether they were selfish or cooperative. They could then choose to punish the other player for the way they had acted. In the first punishment condition, for every dollar you paid, the other person lost three dollars. In the second condition, for every dollar you paid, the other person also lost a dollar.

Around 60 percent of cooperative people chose to punish the selfish player in both conditions. They didn't care about the cost of punishing. However, selfish people behaved very differently. When it cost one dollar to inflict three dollars of damage, 40 percent of selfish people paid to punish. But when it cost one dollar to inflict a one-dollar loss, only 2 percent of selfish people paid to punish. Fairness violations didn't trigger counterdominant behaviors in them. They only acted spitefully when they could improve their status relative to the other player. If punishment couldn't gain them a financial advantage over the other player, they weren't interested.

We can get an idea of the proportion of spiteful acts that are about dominating the other person, as opposed to being about restoring equality. One way to do this is by giving people a choice to punish either to restore equality or to gain a personal advantage. Researchers set up the Dictator Game so that once the dictator made their offer, the other person could pay to destroy some of the dictator's money.[7] The twist was that the person would pay a flat fee of one dollar to destroy as much or as little of the dictator's money as they wanted. Two-thirds of players destroyed an amount of the dictator's money that left them with *more* than the dictator had. They wanted to get ahead of the dictator, not merely to restore equality. Two-thirds of spiteful acts were about getting ahead, about dominating, rather than restoring parity.

Homo rivalis can be seen in other economic games too. In the last chapter, we looked at money-burning experiments. We saw that many people will pay to burn money that others gain undeservingly. But what if other people had more money than you at the end of a game for fair reasons, such as because they worked harder? Would you still punish them?

Researchers investigated that question using the Joy of Destruction game.[8] In it, you are asked to evaluate the quality of some magazine ads. Each assessment takes a good bit of time to do, and you can complete up to three if you wish. You will be paid cash for each ad you evaluate. The researcher tells you that someone else is doing the same task and that, if you want, you can destroy some of the other person's earnings. You don't have to give up any of your earnings to do so (making this "weak spite"), and the other player won't know it was you that destroyed their money. What would you do in this situation? Would you destroy someone else's hard-earned money, just because you could?

Let's start with the good news. If the researcher told participants that the other player would know who destroyed their money, next to no one destroyed money. Fear of retaliation kept spite in check. The vast majority of the spitefully destructive among us are at least spitefully destructive cowards. Unfortunately, as we will discuss in Chapter 7, not everyone is.

Further bad news emerges when we look at how many people chose to destroy some of the other player's money when shielded by anonymity. The number of people who did this was not just predictably depressing; it was shocking. The destruction of other people's fairly earned money was rampant. It occurred in 40 percent of cases.

Why is this figure so high? One possibility is that even fairly earned advantages awaken counterdominant tendencies within us. Recall that in hunter-gatherer tribes today, anyone who tries to big themselves up is brought down. They might have had an exceptionally successful hunt due to their skill and perseverance. No matter. They cannot be allowed to keep most of the food for themselves. Any attempt to do so will trigger counterdominant behaviors from the group. In the last chapter I discussed how anger at unfairness can drive spiteful behavior. However, even if actions are not unfair, other emotions, such as envy (displeasure at another's gain) and schadenfreude (pleasure at another's loss), can also drive us to undertake spiteful actions.[9]

Yet this doesn't seem to be the whole answer. Here's why. Let's say you only evaluated one of the possible three ads. This means you only got paid a third of what you could have earned. You should hence expect that the other player has assessed more ads than you and earned more money. However, if you had evaluated all three ads, the other player could not have more money than you. If counterdominance were driving the destruction of the other player's money, then people who only evaluated one ad should be more likely to destroy the other player's money than people who evaluated all three.

Yet researchers found that players who evaluated only one ad destroyed the same amount of money as people who evaluated three. People weren't simply destroying the other person's money out of counterdominant motives.[10] They seem to have been acting out of dominant motives. After all, if you evaluated three ads, the only reason to destroy the other person's money would be so you could pull ahead of them—to dominate them. As such, we should be careful about essentializing people's character traits, about labeling people categorically either a *Homo rivalis* or a *Homo reciprocans*. The right (or wrong) circumstances can draw dominant behaviors out of nearly half of us.

Justifying your wealth does not keep it safe. Persuading people that you have gotten ahead on merit may not save you from their spite if they can harm you anonymously. In fact, getting ahead on merit, as opposed to luck, might make others feel even more threatened and more likely to spite you. Outcomes signal our status. If another person has more than you due to luck, this doesn't signal that they are better than you. Luck runs out. But if the other person has gotten ahead on their merits, that suggests a stable difference. People are more likely to destroy other people's money if the other person has earned more than them due to merit, as opposed to luck.[11] Talent is more threatening than luck. And, under cover of anonymity, it will get you spited.

THE MORE WE BELIEVE THAT the world is a competition for status, the more dominance pays. This is thanks to the outsized benefits of being number one. We are hence motivated to gain dominance. Because dominant spite can increase our relative position, it should increase as the amount of competition we face increases. Consistent with this, laboratory studies have found that the more competition you face, the more likely you are to act spitefully.[12]

It makes sense that spite should ramp up when we face immediate competitors. If you are competing with the whole world, then it simply matters how much money you have in your wallet. But if you are competing with a couple of other people locally, it matters less how much you have in your wallet and more how much you have in your wallet compared to the others. We display a "positional bias." If given the choice between either having a high income while someone else has a higher income still, and having a lower income that is nonetheless higher than another person's, we tend to choose the lower income. We shun absolute gains for a relative advantage over others.[13] If you are competing in a local group, then acting spitefully to make a relative gain over a rival, albeit while sustaining an absolute loss, can be very helpful.[14]

Studies outside the laboratory have also found that people are more spiteful when they face more local competition. The Nama people of southern Namibia live off their livestock, which graze on commonly managed land. Researchers asked the Nama to play a mini version of the Joy of Destruction game. Nama who had to graze their livestock on poor-quality land were more likely to destroy other Nama's money than Nama who grazed their livestock on high-quality land. For every person on the high-quality land willing to act spitefully, there were two people on the low-quality land willing to act spitefully.[15] When grass is scarce, your fellow citizens are more likely to be viewed as competitors over whom you must prevail. You gain more from harming a competitor when resources are limited.[16]

RESEARCHERS HAVE PROBED HOW THE brain knows that competition is ramping up and hence that it should act more spitefully to get a relative advantage. One way our brain can tell that competition is increasing, at least in the environment in which our ancestors evolved, is when fewer nutrients enter the body, signaling that food is scarce.

An important component of food is tryptophan, an essential amino acid. Our body can't produce it on its own and must import it from food. Without tryptophan, we can't make one of our most important neurotransmitters, serotonin. Drops in serotonin hence signal food scarcity, suggesting that more competition for resources is going on. So spite increases as competition increases, and serotonin levels fall as competition increases. Could it be that the increase in spite is caused by the fall in serotonin? The neuroscientist Molly Crockett and her colleagues have shown this to be the case in a series of remarkable studies.

A 2008 study by Crockett's team invited people into the lab to play the Ultimatum Game, and then to come back a week later and play it again. On each visit, participants were first given a drink. On one visit, it was just a normal drink. But on the other visit, it contained a substance that depleted their tryptophan levels. Crockett found that people acted more spitefully in the Ultimatum Game in the session when they first drank the tryptophan-depleting drink.[17] This suggested that when serotonin levels go in one direction, levels of spite go in the other.

To see if this effect held in both directions, in 2010 Crockett's team tested whether increasing serotonin levels made people less spiteful. They employed the two-drinks procedure again, except this time one of the drinks contained an antidepressant, which increased serotonin levels. As expected, increasing people's serotonin levels made them reject fewer unfair offers in the Ultimatum Game.[18] They acted less spitefully. When serotonin levels go one way, levels of spite do indeed go in the other.

The next question was how serotonin was doing this. Serotonin depletion didn't change people's mood. It didn't alter their tendency to act impulsively. And it didn't make them feel that low offers were more unfair. So why were reductions in serotonin associated with more spite?

In their 2010 study, Crockett's team found that when serotonin levels fell, people became more willing to hurt others. Conversely, when people's levels of serotonin increased, they became less willing to hurt either themselves or others. Crockett's team refined this explanation.[19] They found that reducing serotonin levels increased levels of brain activity in a region called the dorsal striatum when people were punishing others. This region of the brain lights up when we anticipate a reward. Reducing serotonin increases spite because it makes harming others more pleasurable.[20]

The findings from the auction study I discussed in Chapter 1 pushed us toward thinking that spitefulness is typically something we either have or don't. Yet Crockett's work, showing the potential for the environment to influence our levels of spite, suggests we may want to reconsider that conclusion. Spite may be a helpful strategy triggered by encountering a world where it will pay. The harsher the world, the more benefit spite will reap. Consistent with this idea, men who have been exposed to harsh social conditions during childhood, suggesting a world "red in tooth and claw," as Tennyson put it, are more likely to act spitefully.[21]

Serotonin is not the only neurochemical factor influencing levels of spite. In men, testosterone influences spite. Men with higher levels of testosterone are more likely to act spitefully in the Ultimatum Game.[22] This could be because higher levels of testosterone are associated with greater levels of anger.[23] However, it may also relate to the role of testosterone in dominance. As people's testosterone levels increase, they become more concerned with social status. If you give men testosterone, they start to pay more attention to products from the high-status Calvin Klein brand compared to lower-status Levi's products.[24] Testosterone may make them more attuned to matters of social status and more likely to view a low offer in the Ultimatum Game as a threat to that status. They may then react in a way that puts the other in their place, even at a cost. The game changes from social exchange to social competition.

▬▬▬

DOMINANT SPITE CAN HELP YOU excel. This was found by the Greek economist Loukas Balafoutas and his colleagues.[25] They gave people three minutes to add together as many sets of five two-digit numbers (e.g., 10, 76, 45, 23, and 88) as they could. People did this in a group with five other people. In stage 1 they were just told to solve as many sums as they could. In stage 2 they were told there would be a prize for the two people in the group who got the most right.

The study found that spiteful people did just as well as nonspiteful people in the noncompetitive stage 1. But in stage 2, when things got competitive, the spiteful people got significantly more sums right. Competition boosted the performance of spiteful people much more than it did for nonspiteful people. In fact, spiteful people improved twice as much as nonspiteful people. Spiteful people's performance increased by around 30 percent in a competitive situation, whereas nonspiteful people only improved by around 15 percent. Spite made victors too, with 70 percent of spiteful people winning a prize (by being first or second). In comparison, less than half of the nonspiteful people won a prize.

However, Balafoutas and his colleagues also found something paradoxical. When given the chance to choose to compete with others, spiteful people were less likely to decide to compete. Spiteful people were better at competing than nonspiteful people, but had less desire to do so. The reason for this seems to be tucked away in the studies' definition of a spiteful person: hating being behind but loving being ahead (to the extent of being happy to lose money if it meant someone else lost more). Their desire and willingness to be ahead of others, which not everyone shares, helped them excel once plunged into competition. Yet their fear of falling behind made them worried about entering the competitive arena.

We have now seen that both counterdominant and dominant spite can yield direct benefits. This understanding starts to move us away from the "how" of spite toward its "why." What is spite's ultimate explanation? This is what we turn to next.

Four
SPITE, EVOLUTION, AND PUNISHMENT

Different people make different decisions about whether to accept a given offer in the Ultimatum Game. But how much of this difference can be attributed to genetic differences between people? The study of the genetics of spite is still in its infancy. Yet we know there is something woven into our genome that contributes to spite. After asking twins to play the Ultimatum Game, researchers found that 42 percent of the difference between how people respond to offers could be put down to their genes.[1] There is a substantial genetic contribution to spite.

A good guess for how genes do this is through their effects on the dorsolateral prefrontal cortex. We have already seen that this cost-benefit analysis part of the brain contributes to the decision to reject offers in the Ultimatum Game. This part of the brain is under strong genetic control. Other candidate genes are those affecting dopamine levels in the brain, which can influence our emotional responses to low offers.[2] The evolving research adds to our knowledge of the "how" of spite. But let's now turn to the "why." Why did genes for spite evolve?

Think back to the four basic social behaviors: cooperation, selfishness, altruism, and spite. It is obvious why natural selection favors selfishness and cooperation: they directly increase our fitness. But what about altruism and spite? In both cases, our fitness seems to suffer. Altruistic and spiteful genes should therefore be less likely to be passed on to the next generation. So why are altruism and spite still very much here?

This problem was first solved for altruism by one of the twentieth century's leading evolutionary biologists, William Hamilton. Richard Dawkins has described Hamilton as "a good candidate for the title of most distinguished Darwinian since Darwin."[3] Hamilton suggested that it was wrongheaded to focus on how our actions impacted our fitness as individuals. As Dawkins would later put it, we are just vehicles, survival machines, programmed to preserve the genes within us. Hamilton argued that we should assess how our actions impacted the fitness of genes, not the vehicle they happened to reside in. Any gene inside us is likely to have copies of itself in our relatives. We must assess the overall effects of our actions on our genes, wherever they are found. We need to take an inclusive approach.

Like a canny investor, nature has put copies of your genes into many investment vehicles. You have copies of roughly half your genes in your mother, half in your father, half in any children or sibling you have, a quarter in each nephew or niece, and an eighth in each cousin. Thus, if we consider things from a gene's point of view, behaving altruistically toward a group of people who share many of your genes has the potential to benefit your genes, even if it harms you as an individual. As this approach is inclusive, taking into account all the locations a given gene inhabits, it is called "inclusive fitness."

The mathematics of inclusive fitness led biologist J. B. S. Haldane to quip that he would lay his life down for two of his brothers or eight of his cousins. It is somewhat less amusing how this calculus plays out in real life. Ask yourself this: do you think an identical twin would be more likely to sacrifice themselves for their twin or their child? Intuitively, we might think it was their child—after all, who would not put their child above all else? Inclusive fitness suggests otherwise. An identical (monozygotic) twin shares 100 percent of their DNA with their twin but only 50 percent with their child. So what wins in practice: love for your child or the irresistible mathematics of the genome? You have to ask? A 2017 study found being an identical twin predicted a willingness to prioritize your twin over your child.[4] This seemed to be because the twin felt their identity fused with their co-twin. We will return to identity fusion in Chapter 7, where things will get much darker.

Inclusive fitness makes altruism possible. It also makes spite feasible. Altruism is where the personal costs of our actions are outweighed by their benefits to our relatives. Similarly, spite can occur when the personal costs of our actions, when harming another, are outweighed by benefits that flow to our close relatives. This is called "Wilsonian spite," named after the American biologist E. O. Wilson, who framed it this way.

Another way to think of spite is as behaviors that harm my genetic competition more than me. If my spite causes me to lose an arm but causes a nonfamily member who is in close competition with me for a sexual partner to lose their head, then I have benefited. This is called "Hamiltonian spite."[5] Here, the costs of spite flow mainly to people who are less related to me than average and outweigh the personal costs I suffer. Because this spite inflicts costs on genetic competitors, whereas altruism affords benefits to genetic relatives, spite has been dubbed the "shady relative" and "ugly sister" of altruism.[6]

Three conditions need to be met for Hamiltonian spite to evolve. First, your spite must be aimed at your genetic competitors. In the language of biology, it must harm those who are "negatively related" to you. A negative relation is someone who shares fewer genes with you than a random member of the local population. Being spiteful toward negative relations will be genetically beneficial because it will reduce the frequency of competing genes in the gene pool.

To spite your genetic competitors, you need to be able to determine who they are. This is called "kin discrimination." It is the second condition for Hamiltonian spite to evolve, and it is no simple matter. We will touch on it shortly in an example below.

Hamilton thought the benefits from spite would only ever be small. This is partly why biology has overlooked spite for a long time. As a result, Hamilton's third condition for spite to evolve was that it should have little or no cost to the fitness of the creature behaving spitefully. Otherwise, the creature would be out of pocket, evolutionarily speaking. Extinct would be another way to put it.

This final condition gives us a clue about where to begin to search for spite in nature. Some animals can't harm their genetic fitness. One example is a sterile insect. If you are sterile, there isn't much you can do to hurt your reproductive chances because you don't have any to harm in the first place.

The best example of Hamiltonian spite in nature is the sterile worker red fire ant, *Solenopsis invicta*.[7] These ants have a genetic variation in a gene called Gp-9. Worker ants that possess one variant of this gene can smell whether queen ants also have this variant. If a queen doesn't, then the worker attacks her. Within fifteen minutes, the queen will be dead. The worker must be careful though. If it gets the queen's smell on it, other workers will attack it, too. This gene is an example of a "greenbeard." The term comes from a hypothetical scenario popularized by Richard Dawkins in which carriers of an altruism gene have green beards, allowing them to identify each other.

The situation that *Solenopsis invicta* find themselves in meets all three of Hamilton's criteria for the evolution of spite. The worker ants share fewer genes with the queens they kill than they do with a random ant in the population (negative relatedness). Smell provides a simple and obvious way to tell which ants share their variant of the Gp-9 gene (kin discrimination). And there is no direct fitness cost to the workers who kill, because they are sterile.[8]

Sterile wasps are also spiteful.[9] Female polyembryonic parasitoid wasps lay their eggs on the eggs of caterpillars. As a result, the growing caterpillars are eaten from the inside. As the term "polyembryonic" suggests, multiple embryos develop from a single egg, creating genetically identical wasps. Although most wasp larvae grow into normal wasps, some become "soldier morphs." And you thought wasps couldn't get any nastier. Yet wasps must pay a price to become a soldier morph; they become sterile.

Soldier morphs hunt their genetic rivals. They seek and destroy larvae that have developed from other eggs and to which they are thus less genetically related. The benefit of this behavior flows to other wasps who developed from the same egg as the soldier and hence share its genes. They now have fewer genetically dissimilar larvae to compete with.

Even bacteria can act spitefully.[10] Some can harm other bacteria by creating a shower of antibacterial toxins. Doing so comes at a high price. They die. Yet having the gene that enables you to make these toxins also means you are more likely to have another gene that makes you immune to them. So when these bacteria die and release toxins, they preferentially kill genetically dissimilar bacteria. Researchers have begun to consider how we could harness this spite to our advantage. Making bacteria more spiteful could make diseases less virulent.[11] If we can cultivate spite in bacteria, we can create a civil war among them, in which we are the winners.

Spite, conceived of in these Hamiltonian (harm flows to our competition) or Wilsonian (benefits flow to our relatives) ways, can be called "genetic spite."[12] Such spite harms the actor but leads to an overall benefit to their genes. Although I have given some examples here, it turns out to be extraordinarily rare in nature.[13]

This rarity led Hamilton to propose that we investigate spite in nature using a weaker definition of spite. Weak spite, rather than requiring the actor to incur a personal cost while harming another, simply requires the actor not to benefit from harming the other.[14] Such behavior can be found in nature. One example is gulls that destroy their rivals' eggs or kill their rivals' young, but with no immediate benefit.[15]

Another way to weaken the definition of spite is to include acts in which the actor harms both themselves and another in the moment, but personally benefits in the long run. This has been termed "return-benefit spite," "delayed-benefit spite," and "functional spite."[16] Technically such acts are selfish. Yet the immediate act of spitefulness still requires explanation. Examples of such spite are easier to find in the animal world.

An early example was proposed to exist in stumptail macaques in doctoral work performed by the primatologist Alyn Brereton. Brereton found that low-ranked male macaques would intrude on the sexual activities of higher-ranked macaques. There was no direct benefit to the intruder, as they never went on to try to mate with the female themselves. Indeed, it was potentially costly to the intruder, as 3 percent of mating males would attack the intruder. It was also costly to the mating male, as it reduced his chances of successfully impregnating the female macaque. The long-term benefit to the intruder was that they then chose an alternative mating strategy that increased their chances of passing on their genes. Brereton called such behavior return-benefit spite.[17]

Arguments about how spite evolved in humans tend not to focus on inclusive fitness. That is, they don't focus on how spite helps your family or harms your competitors. Instead, as above, they focus on how spite helps you personally in the long run. This spite, which is costly in the moment but can yield personal benefits in the long term, can be called "psychological spite."[18]

An evolutionary model of psychological spite has been put forward by Rufus Johnstone and Redouan Bshary. They explain it as follows.[19] Imagine that you can obtain fitness benefits by attacking and defeating one of your rivals, but there are potential costs associated with the attacks. To get a feel for the costs you might incur in attacking someone, it makes sense to track people's interactions with others. That way, you can try to avoid attacking people who you know are aggressive. Johnstone and Bshary ran the numbers and found that this could lead to the evolution of "occasional spite," in which people will sometimes attack others who are likely to defeat them because doing so gains them a reputation as someone not to be messed with. This is called "negative indirect reciprocity." Other people won't hurt you because they have seen you hurt others. This ties into the reputational benefits of spite that we discussed earlier.

Patrick Forber and Rory Smead have taken another approach. They created a computer simulation to explore what would happen if people used a specific variety of strategies when playing the Ultimatum Game.[20] They designed their study so that each virtual player adopted one of four strategies. Members of one group would accept any offer, and when it was their turn to be the proposer they would make unfair offers (keeping more than 50 percent of the pot for themselves). Let's call this the *Homo economicus* Group. Members of another group were programmed to reject unfair offers and to make unfair offers themselves. We'll call them the Dominant Spite Group. Members of a third group would accept any offer and make fair (fifty-fifty) offers as the proposer. We'll call them the Free-Riding Group, as they aren't prepared to do any costly punishing themselves, leaving that to others. Members of a final group would reject unfair offers but make fair offers. Let's call them the Counterdominant Spite Group.

Now, if everyone begins using the *Homo economicus* strategy, then anyone who changes to one of the other three strategies can't prosper. If you try, you end up worse off than if you had acted like the *Homo economicus* Group. Everyone using the *Homo economicus* strategy is evolutionarily stable. When Smead and Forber ran their simulations, about 70 percent of the time they ended up with people who all used the *Homo economicus* strategy. When they tweaked the conditions of the game, though, things changed.

They introduced what is called "negative assortment," in which a player from one group isn't equally likely to play against every other person. Instead, they are more likely to play against people who don't employ the same strategy they do. Under conditions of negative assortment, Smead and Forber found that it was possible to get a stable population made up of the Dominant Spite and the Free-Riding Groups.

What was interesting was the number of fair offers made if things ended up this way. If there was no negative assortment, spiteful people went extinct and the creatures that remained made fair offers 29 percent of the time. However, with negative assortment, Dominant Spite people could exist and many more fair offers were made. Depending on how extreme the negative assortment was, up to 60 percent of all offers were now fair. Negative assortment allowed spite to exist, and it increased fair behavior.

Smead and Forber's model—and I should stress that it is just a model—suggests that not only can spite evolve but, when it does, it increases the amount of fair behavior in society. The reason the number of fair offers increases is because if you act fairly then spite can't gain a relative advantage over you. If you offer someone five dollars out of a ten-dollar pot and they spitefully reject it, they haven't made any gain on you. You've simply both lost five dollars. Fairness is, therefore, an effective defense against spite.

Smead and Forber's work also has implications for how we think about punishment. Punishment can be understood as harming people who violate norms of behavior, aiming to making them act better in the future. The Counterdominant

Spite Group can be seen as doing this. However, the Dominant Spite Group need not be motivated by punishing unfair behavior. They are just bothered about paying a cost to inflict relative harm. When the Dominant Spite Group forms a stable state with the Free Riders, the situation is not maintained by a threat of punishment, Smead and Forber argue. It is maintained by negative assortment, the Dominant Spite Group being most likely to visit their harm on people who don't employ the same strategy and gaining a relative advantage by doing so.

In considering whether spite played a role in the development of fairness, Forber and Smead invoke the ideas of the German philosopher Friedrich Nietzsche. In his book *On the Genealogy of Morality*, Nietzsche observes that the reason something popped into existence may bear little relation to what it is used for now.[21] He suggests that what punishment is now bears no relation to why it appeared in the first place. Along the same lines, Smead and Forber propose that the behavior we now associate with punishment may have begun as people hurting others to gain a relative advantage over them. Only much later might it have been changed into a mechanism for maintaining fairness and justice. We will return to this idea later in the chapter.

One limitation of Smead and Forber's model is that it simplifies so that people either always or never act spitefully. Naturally, models need to make simplifications. As the Argentinian writer Jorge Luis Borges once observed, the most accurate map is a life-size one.[22] If you think back over your life, have you sometimes acted spitefully and sometimes not?

The variability in spiteful behavior was taken into account in a 2014 computer simulation study led by Xiaojie Chen, from the University of Electronic Science and Technology of China.[23] Chen and colleagues created a model that looked at the effects of people undertaking costly punishment. They did this for the game where everyone first chooses whether or not to contribute to a group investment fund, then receives a payout whether they invested or not. Players can pay to punish those who didn't invest. Those who invest in the group fund are called cooperators, and those who don't are termed defectors. Cooperators who pay to take money off the defectors are called punishers.

Chen and colleagues found that if *none* of the cooperators punished defectors, then the cooperators could just about survive. But to do so, they had to bunch together in enclaves. In contrast, if *all* the cooperators punished defectors, the cooperators died out. The cost of punishment was too high for them. What was remarkable was what happened if the cooperators punished only half the time: all defectors were eliminated. This seems odd. Punishing half the time means combining two losing strategies. If you never punish you tend to lose, and if you always punish you definitely lose. Yet alternating between the two losing strategies led to triumph. This is an example of Parrondo's paradox: combining two losing strategies can sometimes be a winning strategy.

The question then becomes: why, in the real world, would we sometimes undertake costly punishment and sometimes not? The authors propose that the answer lies in our emotions. The emotion of anger can be unpredictable.

Sometimes we lose our cool, other times we don't. Such unpredictability may be a virtue, not a fault.

———

WHAT IS COSTLY PUNISHMENT OF unfairness really about? Do we do it to deter people, to make them behave better in the future? Or do we do it to retaliate, to reduce the status and competitiveness of the other person, meaning punishment has a "competitive function"? In an intriguing 2019 paper, Nichola Raihani and Redouan Bshary argue that we have overestimated how much punishment is about increasing cooperation. Instead, they propose that it is much more about retaliation.[24]

Raihani and Bshary point out that there are a series of problems associated with the idea that punishment is a tool for encouraging cooperation. First, we punish people who are already being cooperative. This is the do-gooder derogation we saw earlier. Such punishment reduces, rather than increases, cooperation. However, this isn't a conclusive argument. Defibrillators can be used to hit people over the head, but that doesn't mean bludgeoning is what they were designed for.

Another problem is that if punishment were about promoting cooperation, we should simply punish unfair people. How much the other person's unfairness caused them to gain on us personally shouldn't impact our decision. However, as Raihani and McAuliffe have found, this is not the case.[25] Their experiment involved a two-player game where one player could steal money from the other. The victim always suffered the same loss. However, the relative outcome varied. Sometimes the thief still ended up with less cash than their victim, sometimes they had the same as the victim, and sometimes they had more. The study found that the victim's decision to punish the thief was strongly influenced by whether the thief ended up better off than the victim. People punished when they were made worse off by another person breaking a rule. They did not simply punish because the other broke a moral rule. Their punishment was not about increasing cooperation. It was about harming someone who had unfairly advanced ahead of them.

There are also questions over the evidence held up to show that punishment increases cooperation. Although studies have found that punishment is associated with greater levels of cooperation,[26] this doesn't prove that punishment caused the increase. People may have changed their behavior because they realized the other person was willing to cooperate or that cooperation was the norm of the group.[27] More importantly, such studies often fail to reflect the real world. They don't allow the potential for retaliation or do-gooder derogation. They ignore that we rarely engage in "one-off" interactions with others. Yet, in life, there is almost always another round.[28]

When you make economic games and simulations more like the real world, the evidence that punishment increases cooperation weakens. If you allow do-gooder derogation in computer simulations, then letting people punish others leads to a negligible increase in cooperation.[29] If you allow people to pay to punish others while playing the same game repeatedly, it increases levels of cooperation but not the amount of money the players make. In fact, those who make the most money are the ones who use costly punishment least. Winners don't punish. The researchers who reached this conclusion suggest that costly punishment might have evolved for reasons other than promoting cooperation. They propose that spite evolved as a way of "coercing individuals into submission and establishing dominance hierarchies."[30]

The timing of punishment in economic games also makes us suspect that punishment is often used to dominate rather than reform. Imagine you are playing a multiround game with someone. If you aim to use punishment to make them play fairer with you, it will make no sense for you to exact the greatest punishment in the final round. After that ultimate punishment, you will never see them again. Yet this is when most punishment is inflicted, which looks like people capitalizing on their final chance to hurt someone rather than acting to reform the other's behavior. It looks like the bitterness often seen in a divorce, the final round of a marriage. Such behavior is consistent with punishment having a harm-focused, competitive aim, rather than trying to deter unfair behavior.

Furthermore, punishment occurs in other situations where it can't possibly deter cheating. For example, Crockett and colleagues tried to tease apart the retaliatory and deterrence functions of punishment.[31] They found that there was a 15 percent chance that people would pay to punish someone who had screwed them, even when this other person would not know they had been punished. Here, people were punishing in order to harm the other, not for any didactic purpose. If told the other person would know they had been punished, there was now a 20 percent chance that people would pay to punish them. This small increase could represent the punisher trying to deter the other person from acting unfairly. If so, most of the punishment would serve a retaliatory purpose, but some would be for deterrence. However, another interpretation of this 5 percent rise is that the punisher was trying to inflict emotional damage on the other (because the other would know that the punisher was angry with them), in addition to financial harm. Thus, what may appear like a small attempt at deterrence could have been a further ramping up of retaliation.

People don't even know they are acting out of retaliation when they clearly are. Again, we know this thanks to the work of Crockett and colleagues. They were able to get a measure, derived from people's behavior, of how much their punishment was driven by deterrence. An act of deterrence is one that aims to teach someone a lesson, by letting them know they have violated a norm, to try to make them behave better in the future. Crockett's team also measured how much participants' punishment behavior was driven by retaliation (simply aiming to inflict damage to make the other suffer).[32] When they asked people whether they were acting out of deterrence, people's answers matched up with what their behavior showed they were doing. Yet when they asked people about the extent to which their actions were driven by retaliation, people's answers bore no relation to their actual behavior. They were retaliating but didn't know it. This makes some sense because, as we saw earlier, people don't like it when others take personal revenge. There is hence pressure on us to hide our retaliation, even from ourselves, behind a screen of decency. We delude ourselves into thinking we are punishing for noble purposes. In reality, we are concerned about harm and status.

Crockett and colleagues' work is consistent with the ideas of Smead and Forber. Here, punishment doesn't evolve because it encourages noncooperators to reform themselves. Increases in cooperation and fairness are just side effects of people being willing to pay to hurt others and gain a relative status advantage. In this view, we first evolved the ability to spite others to gain relative status, then repurposed this tendency into punishment.

We often intuit this competitive aim of punishment. Recall the study discussed earlier in which people could write a note when deciding whether to accept or reject an offer in the Ultimatum Game. One player, when rejecting the offer, wrote a note stating:

> Sorry, I'm a person too. When the cards are all in my hand, you should try to appease me instead of offend me. There was a 50/50 split. It couldn't have been easier. So, since you decided you are obviously better than I am. You get nothing. Enjoy it, I know I will.[33]

Why we punish matters. If we are driven by a desire to inflict suffering and gain status, rather than to reform the other person's behavior, we will create dangerous social systems that achieve the questionable end we seek.[34]

In a competitive world, spite emerges as a potentially adaptive strategy for getting ahead of your local competition. The strong social pressure not to be seen as spiteful makes us wrap the wolf of spite in sheep's clothing. The best way to make others believe that your spiteful attempt to obtain a relative gain is really for the greater good is to believe the lie yourself. Out of this attempt to dominate comes an accidental benefit: more acts of fairness in society. A vice has led to virtue.

This is not the only accidental benefit of spite. Spite was meant to be a way to deal with people. Yet it can also take on more abstract enemies. This, too, can yield surprising benefits.

Five
SPITE AND FREEDOM

eason dictates what we ought to do to survive and flour-ish. But we have to neither like it nor obey it. We are ra-tional creatures that can rage at reason. We can yearn for something else. As the former Harvard psychologist, now spiritual teacher, Ram Dass puts it, we would rather be free than right.

Do we target our spite only at people? Or would we dare blaspheme against the Enlightenment and spite reason it-self? The French philosopher Jean-Jacques Rousseau thought we wouldn't. "The nature of things does not madden us," he wrote, "only ill will does."[1] But Captain Ahab disagrees.

"Talk not to me of blasphemy, man," he roars in our ears. "I'd strike the sun if it insulted me." If we were indeed prepared to lash out to spite logic, the laws of nature, and inevitability itself, this would make us tragic beasts. Yet, as we will see, such apparently nonsensical behavior can also have benefits. We are at least gloriously tragic.

The solutions to the formidable global problems we face, argues another Harvard psychologist, Steven Pinker, lie in reason.[2] This is consistent with the Enlightenment principle that we must use reason to understand our world and overcome our follies. After dismissing faith, authority, and gut feelings as "generators of delusions," Pinker argues that the use of reason when making decisions is "non-negotiable."

Unfortunately, trying to tell people they must do something can backfire. In the early 1960s, the Yale psychologist Stanley Milgram undertook a series of studies into obedience. A scientist instructed volunteers to give increasingly severe electric shocks to a fellow volunteer in another room, ostensibly to help them learn.[3] Sixty-five percent of volunteers were still administering shocks when they reached the maximum possible 450-volt shock.[4] Yet a 2009 partial replication of this study found something strange. When the scientist told volunteers, who were wavering about whether or not to obey, "You have no other choice, you must go on," all of them chose to disobey.[5] One explanation for this reaction starts with self-determination theory. This proposes that we have a basic psychological need for autonomy, a need to feel in control of our fate, a need to have "a feeling of choice."[6]

First used to describe Greek city-states employing their own laws, the term "autonomy" comes from the Greek words *autós* ("self") and *nomos* ("law" or "rule").[7] When applied to human beings, it yields the concept of the autonomous individual. This creature was an invention of the Enlightenment.[8] The autonomous individual is not free from external influences but can thoughtfully assent to or dissent from them.[9] They are bound to no laws but those of their own making.[10] They can use reason and rational thought to guide their behavior according to their own values.

The extent to which individuals are autonomous is unclear. Beneath the tranquil simplicity of the concept of autonomy swim profound and unresolved problems.[11] To what extent are our values genuinely our own? Is it always problematic to constrain people for their own good? Does the idea of a "self" that chooses make sense? Conceiving of and treating people as autonomous is the worst way to approach them, except for all the others.[12]

Our need for autonomy can be measured using the concept of reactance.[13] Reactance is how strongly an individual reacts if their freedom is impaired. Patrick Henry, who famously cried, "Give me liberty or give me death," would score highly on reactance. Reactance varies from person to person and across the life span. As any parent knows, it rears up in both the "terrible twos" and adolescence. As any child knows, it can reappear in parents' senior years.[14]

But why do we have a feeling of autonomy to be threatened in the first place? To answer this, consider what would happen if we had no sense of free will. Would we no longer feel responsible for our actions? To paraphrase Dostoyevsky, if free will did not exist, then are all things permitted? This idea was tested in a 2008 study by Kathleen Vohs and Jonathan Schooler in which participants read passages from a book by the Nobel laureate Francis Crick.[15] One group read Crick's claim that smart people now believe that free will is an illusion. Another group read a passage from Crick's book that had nothing to do with free will. The researchers then gave both groups a math test in which they had an opportunity to cheat. The group exposed to Crick's arguments against free will were more likely to cheat. Vohs and Schooler conclude that losing belief in free will could encourage socially undesirable behavior. No free will, no responsibility, no worries.

The following year, another group of researchers found further evidence consistent with this idea. The American psychologist Roy Baumeister and colleagues asked one group of people to read sentences that supported free will and another group to read sentences that argued against it.[16] The group induced to believe less in free will became less willing to help others and more willing to act aggressively. This is worrying given that large segments of the population increasingly feel less in control of their lives.[17] The extent to which that sentiment may be affecting the social fabric of our world is not yet clear.

Whether or not we have free will, most of us feel we do. And we guard it jealously. Threatening our sense of freedom triggers what I call the Braveheart Effect.[18] This is the motivation that rushes up in us to restore our lost or threatened freedom. We can all remember the moment in the film *Braveheart*, starring Mel Gibson, in which William Wallace bellows, "They may take our lives, but they'll never take our freedom!" Wallace then charges into battle to attempt to regain the freedom that the English had stolen.

The Braveheart Effect begins with rising anger and hostility.[19] Negative thoughts enter our mind about the person or group that has threatened our freedom. We begin to dislike what we were forced into. We want what we were denied. Next we act to try to regain our lost sense of freedom. Sometimes we do this by doing what we have been explicitly told not to do. If a judge instructs a jury to ignore what they have just heard, the jury may place more weight on the inadmissible events, not less.[20] The Braveheart Effect can also lead us to do the opposite of what someone predicts we will do.[21] Additionally, being told you have no choice but to believe something that you happen to believe already can make you believe it less.[22]

There are clear links between the Braveheart Effect and spite. When we are confronted with a threat to our freedom, the Braveheart Effect can trigger a spiteful response. We will pay a price to punish whoever or whatever has taken our liberty, in an attempt to regain that sense of freedom. It may be another person who has threatened our freedom. It could be a country. Or it may be the laws of nature and reason that have restricted us. Let's look closer at spite directed against abstract entities. Whereas the Enlightenment has encouraged us to view reason and freedom as going hand in hand, under certain conditions they may have their hands around each other's throats.

IN TOTALITARIAN STATES, REASON CAN be a tool of liberation. In George Orwell's *Nineteen Eighty-Four*, Winston Smith is tortured by an agent of the state to make him say, believe, and perceive that two plus two equals five. As one critic observes, "When Orwell's hero is fighting for 'two plus two is four' and repeats this over and over again as a secret formula for life and freedom . . . [it is] the symbol of freedom, freedom from manipulation by the omnipotent party."[23]

Yet other authors have represented the brute fact of "two plus two is four" as potentially oppressive. Fyodor Dostoyevsky dramatized this in his novella *Notes from the Underground*, which begins with the scene-setting line "I am a sick man. . . . I am a spiteful man."[24] The character of the Underground Man explains:

> Two times two makes four seems to me simply a piece of insolence. Two times two makes four is a fop standing with arms akimbo barring your path and spitting. I admit that two times two makes four is an excellent thing, but if we are going to praise everything, two times two makes five is also a very charming little thing. . . . Man is a frivolous and incongruous creature and perhaps, like a chess player, loves only the process of the game, not the end of it.

For the Underground Man, this simple sum is a "symbol of human unfreedom in relation to the laws of nature."[25] He experiences as oppressive the idea that the demands of reason should drive human behavior. The Braveheart Effect kicks in. He acts to restore his feeling of freedom. He does so by acting spitefully. His spite aims at reason itself. He refuses to see a doctor about his liver. He takes pleasure in toothache. He revels in his degradation. The Underground Man knows he cannot overcome the laws of nature, but he also knows he does not need to like them: "I cannot break through the wall by battering my head against it if I really have not the strength to knock it down, but I am not going to be reconciled to it simply because it is a stone wall and I have not the strength."

Dostoyevsky has the Underground Man put forward a view of human nature in which people are driven not by rational self-interest but by the need to feel free. As he states it, "What has made them conceive that man must want a rationally advantageous choice? What man wants is simply *independent* choice, whatever that independence may cost and wherever it may lead. And choice, of course, the devil only knows what choice."

The ideas in Dostoyevsky's book, published in 1864, arose from the turbulence in Russia at the time.[26] At the beginning of 1861, a third of the Russian population were serfs, unfree peasants tied to the land of their landlords. By the end of it, none were. In February of 1861, Emperor Alexander II signed statutes into law that abolished serfdom. Twenty-two million men, women, and children received their freedom.[27] This provided the impetus for Russian intellectuals to search for a new type of human.[28]

At the time, there was a generational split in Russia, famously portrayed by Ivan Turgenev in his novel *Fathers and Sons*. An older generation, the fathers of Turgenev's title, were men of the 1840s. They were a cohort of romantics and idealists who sought to create liberty, solidarity, and equality by building on the natural goodness of humans. Their sons, men of the 1860s, rejected the romanticism of their forebears. Instead, they worshipped at the shrine of the European Enlightenment. For them, radical change needed to come through reason and science, and it needed to happen immediately.

The new men of the 1860s had a mechanistic understanding of humanity. They subscribed to the doctrine of naturalism. Darwin's *On the Origin of Species*, published in 1859, encouraged a view of humans as merely another physical object in the world and subject to the laws of nature. Naturalism proposed that people should be studied with the same scientific detachment as we study animals or rocks. Turgenev captured this in his portrayal of one of the men of the 1860s, Bazarov, when he was dissecting a frog. Bazarov explains, "I shall cut the frog open and see what's going on in his insides and then, as you and I are much the same as frogs, only that we walk on legs, I shall know what's going on inside us, too."

These young men were utilitarians. They viewed the purpose of life as maximizing happiness. What was right was, therefore, whatever maximized happiness, and all rational actions should aim for that end. When this generation combined naturalism and utilitarianism, humans became machines that sought to gain pleasure and avoid pain.

The Russian writer Nikolai Chernyshevsky portrayed the creature that resulted from this ideology in his smash-hit 1863 book, *What Is to Be Done?*, subtitled *Tales About New People*. In it, Chernyshevsky glorified humans who were dedicated to following the dictates of reason and science. In Chernyshevsky's view, ignorance caused sin. If people simply understood their interests, they would necessarily act to achieve them, becoming citizens of a perfectly moral and rational society. The future was to be "bright and beautiful."

Dostoyevsky thought that Chernyshevsky was deluded. If *Notes from the Underground* had been published today, it would likely have been retitled *The Chernyshevsky Delusion*. When Dostoyevsky looked at humans, he did not see creatures that would automatically act in their best interests when they were told what those were. He saw a creature that could know the light yet still run off into the black and dangerous night. To think otherwise was to deny a stubborn, dark side of human nature. Something comparable to this sentiment can be found today in the work of the philosopher Nicholas Taleb. He writes, "If you are not a washing machine or a cuckoo clock—in other words, if you are alive—something deep in your soul likes a certain measure of randomness and disorder."[29]

The difference between Dostoyevsky and Chernyshevsky foreshadowed the ideas of the American writer Thomas Sowell, as put forth in his book *A Conflict of Visions*.[30] Sowell argued that there are two basic stances we can take toward human beings. The first is the Unconstrained Vision. Here, people are naturally good and can become anything. We are perfectible creatures. In this vision, the problem of evil can be solved by education, reason, and changing people's environments. The alternative is the Constrained Vision. Here, people are flawed by their very nature. Evil is inevitable and must be constrained. Authority and tradition are needed to control humanity's dark side. We dismiss them at our peril.

Dostoyevsky saw the potential for people to feel trapped, not liberated, by naturalism and utilitarianism. As he has the Underground Man say:

> As soon as they prove to you, for instance, that you are descended from a monkey, then it is no use scowling, accept it for a fact. When they prove to you that in reality one drop of your own fat must be dearer to you than a hundred thousand of your fellow-creatures and that this conclusion is the final solution of all so-called virtues and duties and all such prejudices and fancies, then you have just to accept it, there is no help for it, for twice two is a law of mathematics. Just try refuting it.

Reason was not a bad thing, but it was only a part of our nature. In the words of the Underground Man, "Reason is an excellent thing, there's no disputing that, but reason is nothing but reason and satisfies only the rational side of man's nature, while will is a manifestation of the whole life." Instead, Dostoyevsky believed that there was something more precious to people than pleasure and reason. As his Underground Man explains, "The fact is, gentlemen, it seems there must really exist something that is dearer to almost every man than his greatest advantages . . . for the sake of which a man if necessary is ready to act in opposition to all laws; that is, in opposition to reason, honour, peace, prosperity."

Dostoyevsky claimed that we had a compelling need to feel free. People would therefore be likely to reject the idea that they were machines running according to laws. They would refuse to be pianos played by a celestial pianist. The Underground Man argues that even if we were showered with every pleasure, comfort, and blessing,

> out of sheer ingratitude, sheer spite, man would play you some nasty trick. He would even risk his cakes and would deliberately desire the most fatal rubbish, the most uneconomical absurdity . . . simply in order to prove to himself—as though that were so necessary—that men still are men and not the keys of a piano, which the laws of nature threaten to control so completely that soon one will be able to desire nothing but by the calendar.

Even if science could show that humans' behavior was completely determined, this would, in the Underground Man's opinion, still lead some people to

> purposely do something perverse out of simple ingratitude, simply to gain his point. And if he does not find means he will contrive destruction and chaos, will contrive sufferings of all sorts, only to gain his point! He will launch a curse upon the world and as only man can curse (it is his privilege, the primary distinction between him and other animals), may be by his curse alone he will attain his object—that is, convince himself that he is a man and not a piano-key! If you say that all this, too, can be calculated and tabulated—chaos and darkness and curses, so that the mere possibility

of calculating it all beforehand would stop it all and reason
would reassert itself, then man would purposely go mad in
order to be rid of reason and gain his point! I believe in it, I
answer for it, for the whole work of man really seems to con-
sist in nothing but proving to himself every minute that he is
a man and not a piano-key! It may be at the cost of his skin.[31]

In this view, it is less that we are rational beings that refuse to be rational and more that we are machines that refuse
to be machines. The Underground Man is prepared to spite
reason and inevitability to shake his cogs from their moor-
ings and feel free. Acting against our rational best interests,
defying reason because we get the payoff of feeling free, is
what I call "existential spite."

———

FOR EXISTENTIAL SPITE TO EXIST, we must find the feeling
of freedom to be intrinsically rewarding. We need to desire
it as an end, not simply as a means. There has been much debate over whether freedom is of instrumental value (i.e., we
value it because it helps us achieve other things) or intrinsic
value (i.e., we value it for its own sake).[32] Yet it makes sense
that freedom should provide both types of value. Sex enables
us to have children, but we also value it for its own sake.

It would be simple enough, in our modern world, for us
to feel our freedom threatened by things other than people
physically trying to gain power over us. The products of the
Enlightenment, whether they be newly discovered laws of
nature or the prescriptions and warnings of the new expert
class it yielded, can be perceived as restricting our freedom.

Spiteful rebellion could ensue. Counterdominant spite could lead us to try to get one over on inevitability.

Trying to resist attempts at domination by other people makes some sense. Yet applying the same sentiment to reason seems to be not only foolhardy but positively dangerous. The decoupling of spite from its interpersonal context and its application to an abstract realm of forces could have severe consequences. We would be like moths, designed to fly toward the light of the moon, but now drawn to electrified lamps. Existential spite would lead humans to be defined not only by their rationality but also by their willingness to oppose it.[33]

Existential spite, paying a cost to retain a feeling of freedom in the face of the dictates of reason, would appear to have only downsides. When it comes to the laws of nature, the house always wins. Existential spite could only end in calamity. Or would it? There are two objections to this assertion.

First, it can be rational to refuse reason. Steven Pinker's masterful ode to reason, *Enlightenment Now*, claims reason as the only solution to humanity's problems. I don't doubt that he is right, in the long run. But reason has a dark side. It is the only acceptable form of domination left in our society. The better-reasoned argument wins through what the philosopher Jürgen Habermas calls "unforced force" (*zwanglose Zwang*).[34] Yet unforced force is still force. People resent domination by reason. Michael Gove, the former UK Justice Secretary, captured this sentiment during the 2016 Brexit Referendum Campaign when he said, "People in this country have had enough of experts."[35]

One could argue that if others try to dominate you by reason, then reason can also act as its own counterdominance device. If someone tries to reason you into something, you can try to reason your way out of it. Yet this claim presupposes that everyone has equal access to reasoning. Reasoning takes resources. As Virginia Woolf once put it, "Five hundred [pounds] a year stands for the power to contemplate." Reasoning takes space. To quote Woolf again, a "lock on the door means the power to think for oneself." And reasoning takes time. I don't have a Woolf quote for this one. Either way, reason has the potential to be used as a tool of domination by the privileged. In the face of advanced sophistry or reasoning that one does not have the privilege to unpick, spite may be an appropriate counterdominant response.

Reason is also hubristic. The Enlightenment thinker believes that their reasoning is superior to the accumulated millennia of tried and tested knowledge contained in tradition. They may be right. But they have often been wrong. Nevertheless, we are taught to laud the progressive reasoner and demonize the regressive traditionalist. We can link this back to Sowell's two visions. Those with an Unconstrained Vision believe we can reason ourselves to utopia. In contrast, those with a Constrained Vision are more skeptical and think we need to rely to a greater degree on established tradition.

The work of Joseph Henrich, whose anthropological work we encountered earlier, gives evolutionary reasons to be concerned about reason. In his remarkable book *The Secret of Our Success*, Henrich first documents examples of individual reason being inferior to accumulated tradition.[36]

He discusses European explorers who visited places where hunter-gatherers had survived for centuries. Without the benefit of tradition to guide them, being armed only with their wits, these explorers often starved to death.

Not only can reason be inadequate, claims Henrich; it can also be positively dangerous. He documents numerous situations where tradition will keep you safe, but reason will get you killed. Take the case of pregnant women in Fiji. They face a cultural taboo against eating shark meat. When asked why they didn't eat shark, women said they thought it was because their babies would be born with rough skin. If they used reason, they would see this was ridiculous. They might then go on to eat shark—which would be a mistake. Tradition is wiser than reason here. Eating shark while pregnant can damage the unborn child due to the toxins that can accumulate in the flesh of the toothy fish.

If, historically, an individual's efforts at reasoning were likely to prove dangerous, by undermining the wisdom of tradition and threatening to get one killed, spiting reason might not be a bad idea. Sometimes. Today, our ability to reason is improved. We have new Bayesian approaches[37] and systems such as peer review that help our thinking. This fact leaves existential spite as a tool to combat domination by reason rather than a way to escape the potential death trap of reason.

A second potential benefit of existential spite is that it makes people more creative. Cultivating spite in the face of impossible odds could encourage us to persevere and create new solutions to the problems we face. Knowledge liberates by preserving us from the frustration of attempting the impossible, wrote Isaiah Berlin.[38] But it's from "the champions of the impossible, rather than from slaves of the possible, that evolution draws its creative force," argues Barbara Wootton.[39] In this vein, we can think about existential spite in relation to goal setting.

When I used to work in the world of finance, it was banged into me that goals should be SMART: specific, measurable, achievable, realistic, and timely. This remains standard advice today. But should goals be realistic? Should we sometimes set goals that others tell us are not realistic? We could use the Braveheart Effect to create reactance to motivate us to achieve such goals. In essence, we would embrace a spiteful response to the "it can't be done" sentiment. We would expose ourselves to a high risk of failure and try to inflict a cost on reason and conventional wisdom by proving them wrong. Such acts are spiteful in the moment, although one would hope to see delayed benefits from such efforts, as no one sets out to fail.

Something like this can be found in the world of business, in the concept of the "stretch goal": one that is considered virtually unattainable or seemingly impossible.[40] The first ambitious project reconceived of as being a stretch goal was President John F. Kennedy's aim to land a human on the moon by the end of the 1960s, something not considered technically feasible at the time.[41] Such goals, characterized by extreme difficulty and tremendous novelty, are hence sometimes called "management moon shots."[42]

Spitefully going against received wisdom to achieve the impossible seems to be deeply ingrained in us. Part of happiness, Nietzsche argued, comes from "the feeling that power increases." Nietzsche meant to highlight the pleasure we get from overcoming resistance. For him, the pharmacological happiness experienced by the population in Aldous Huxley's *Brave New World* would have been hell.

George Orwell also noted this need for struggle. In his 1940 book review of Hitler's *Mein Kampf*, Orwell observed that nearly all Western thought had assumed that people "desire nothing beyond ease, security and avoidance of pain." This was a strictly utilitarian view of life. Orwell recognized "the falsity of the hedonistic attitude to life."[43] There was, he noted, more to life than this.

Jesus and Hitler wouldn't have agreed on much. Yet there was one belief they shared: humans cannot live on bread alone. Hitler, as Orwell observed, knew that people "don't only want comfort, safety, short working-hours, hygiene, birth-control and, in general, common sense." Orwell credits Hitler with the insight that people, "at least intermittently, want struggle and self-sacrifice, not to mention drums, flags and loyalty-parades." In this way, wrote Orwell, "Fascism and Nazism are psychologically far sounder than any hedonistic conception of life." Orwell continued:

Whereas Socialism and even capitalism in a more grudging way, have said to people "I offer you a good time," Hitler has said to them "I offer you struggle, danger and death," and as a result a whole nation flings itself at his feet. Perhaps later on they will get sick of it and change their minds, as at the end of the last war. After a few years of slaughter and starvation "Greatest happiness of the greatest number" is a good slogan, but at this moment "Better an end with horror than a horror without end" is a winner. . . . We ought not to underrate its emotional appeal.

Perhaps the most eloquent example of our desire for struggle, and one that thankfully does not invoke Hitler, comes from the English science fiction writer Douglas Adams. In the BBC radio version of *The Hitchhiker's Guide to the Galaxy*, Adams places his protagonist, Arthur Dent, on a planet in which medicine has advanced to the point that it can cure or heal practically anything. Yet this has an unexpected effect:

Like most forms of medical treatment, total cures had a lot of unpleasant side effects. Boredom, listlessness, lack of—well anything very much, and with these conditions came the realisation that nothing turned, say, a slightly talented musician, into a towering genius faster than the problem of encroaching deafness. And nothing turned a perfectly normal, healthy individual into a great political or military leader better than irreversible brain damage. Suddenly everything changed. . . . Doctors were back in business—recreating all the diseases and injuries they had abolished—in popular, easy-to-use forms. Thus, given the right and instantly available types of disability, even something as simple as turning on the 3D TV could become a major challenge. And when all the programs on all the channels actually were made by actors with cleft-palates, speaking lines by dyslexic writers, filmed by blind cameramen, instead of merely seeming like that, it somehow made the whole thing more worthwhile.[44]

Adams is effectively talking about stretch goals: self-made crises that try to spur feelings of meaningfulness through struggle.[45] Existential spite can push us into working to achieve stretch goals, activating our deep desire for struggle and thereby creating new, innovative, and unimagined solutions to problems. Indeed, to attempt stretch goals is to mandate creativity.[46] As Steve Kerr, a former chief learning officer and managing director at Goldman Sachs, highlights, stretch goals can get "your people to perform in ways they never imagined possible."[47] Existential spite can produce breakthroughs.

As you can imagine, stretch goals can also go horribly wrong. Employees may find them demotivating, particularly in companies with an intolerance for failure.[48] They can lead to learned helplessness: experiences of defeat and depression in the face of problems we cannot control or escape. Organizations may also run into trouble by setting such goals. In 2001 the European carmaker Opel was struggling. That year it lost more than half a billion dollars. With limited resources, it attempted a stretch goal: to return to profitability in two years. It did not come close, and the failure only deepened morale problems at the company.[49]

Medhanie Gaim and colleagues argue that the carmaker Volkswagen's objective of producing cars that were fast, cheap, and green was another example of a stretch goal gone wrong.[50] The aim of achieving performance, efficiency, and cleanliness seemed unattainable. More efficient diesel engines came with higher emissions, and better performance meant lower fuel efficiency. Other German carmakers, such as BMW and Mercedes-Benz, had determined that the "fast-cheap-green" goal was effectively impossible. In the end, Volkswagen could only accomplish it by installing a gadget that turned on emission controls only during emission testing. By early 2019, the scandal had cost them $30 billion.

Despite their potential for calamity, stretch goals can lead to great success. In 1972, Southwest Airlines faced the problem of having only three rather than four airplanes and needing to do more with what it had. As a result, the company set a goal of turning around its planes at the airport gates in just ten minutes. Pretty much everyone, including its competitors, the US Federal Aviation Administration, Boeing, and some of its own employees, thought this was impossible. Back then, the typical time for turnaround was an hour. Yet by making radical changes, including drawing on the behaviors of racing-car pit crews, Southwest was able to achieve its goal.[51]

Another example of a company successfully setting and achieving a stretch goal comes from DaVita, which provides kidney care. As Sim Sitkin, of Duke University's School of Business, and his colleagues explain, DaVita faced the problem that 90 percent of its patients were in government programs that didn't pay the full cost of treatment. In response, DaVita decided to set a goal of creating a ridiculously large ($60–$80 million) cost-savings within four years, while simultaneously improving both patient outcomes and employee satisfaction. To do this, the company created a new unit, the Pioneer Team. The team leader, Rebecca Griggs, recalls, "We had no idea how to achieve such large savings in such a short time frame. In fact, we had no idea if it was even possible." But within a few years, her team had already saved $60 million, reduced the number of patients who were hospitalized, and increased employee satisfaction.[52]

The question then becomes: what factors influence whether a company will succeed or fail at a stretch goal? Sitkin and his colleagues argue that companies should undertake stretch goals when they are already well positioned and on a winning streak. A company that attempts stretch goals when it is weak communicates fear and desperation. Unfortunately, this is when management may be most likely to attempt them. To make this point, Sitkin draws on the psychological literature on loss aversion and decision-making. Psychologists Daniel Kahneman and Amos Tversky famously showed that failure makes people more inclined to take risks to dig themselves out of a hole.[53] As a result, struggling firms are more likely to take risky actions.

The other implication of Kahneman and Tversky's work is that successful firms are likely to be more risk averse, despite the fact that they are the ones with the resources and motivation to fruitfully take risks and achieve stretch goals. Such companies can become complacent. One way to get around this is to artificially make companies feel threatened, similar to the scenario described in the above quote from Douglas Adams. Sitkin describes how Kenneth Frazier, CEO of the pharmaceutical company Merck & Co., used this approach.[54] Frazier asked his executives to imagine that they were Merck's competitors and to brainstorm what they would do to beat Merck. This created a sense of risk, which motivated them to take on new stretch goals.

Stretch goals can also be used to attempt to solve the environmental problems we face. This begins with stakeholders imagining "provocative alternative futures" and then setting goals to reach their vision. For example, the Scottish Trees for Life project aims to restore forests in the north-central Scottish Highlands. This includes reintroducing beavers, wild boar, lynx, moose, brown bears, and wolves. Many would deem such an aim unrealistic, yet it is also an inspirational and guiding vision for shorter-term milestones.[55]

Whether one is in the business of business or of saving the planet—or indeed involved in any form of creative activity—setting what seem like impossible goals may be an effective way to trigger the Braveheart Effect and utilize existential spite. By artificially summoning our counterdominant side to take on domineering abstractions, whether they be beasts of reason or tradition, spite can be used to advance the world and achieve the "impossible."

Revolts against injustices that have been imposed by overwhelmingly powerful elites and tyrants can also make use of existential spite. Change in the face of such forces may seem hopeless. It may be thought madness to even try. Yet sometimes, madness is the only appropriate response. For the American theologian Reinhold Niebuhr, only a "sublime madness in the soul" would make people "battle with malignant power and 'spiritual wickedness in high places.'"[56] Liberalism, Niebuhr felt, was hindered by "the gray spirit of compromise." It lacked the fanaticism to "move the world out of its beaten tracks."[57] Existential spite may be a tool that can help summon sublime madness.

Yet, as Niebuhr stressed, sublime madness could also encourage terrible fanaticism. We will come to this in Chapter 7. What Niebuhr warned of in relation to sublime madness also applies to existential spite. Both must some-how be "brought under the control of reason." We can only hope, Niebuhr added, that reason will not destroy sublime madness before its work is done.[58] This assumes that the practitioner of existential spite will remember to recall reason. It is a perilous balancing act, and the abyss below is all too menacing.

In such ways, our biological drive to spite can be used to create new and positive cultural change. Yet spite can also have wider cultural impacts. How it can do so is an important question, no more so than in the realm of the political.

Six
SPITE AND POLITICS

lections are the perfect places for spite to manifest. Politics is where we debate what is fair. Perceived unfairness and others getting ahead of us cause anger, the essential political emotion.[1] This can in turn trigger counterdominant spite. In the political realm, as we will see in this chapter, counterdominant spite can take the form of a "need for chaos," with potentially apocalyptic effects. Politics is also where we try to get ahead. We may employ dominant spite to this end. A fear of retaliation would normally restrain spite, but in elections we disappear behind a curtain to vote. In the darkness of anonymity, spite slips its chains. What

135

we have learned about spite can contribute to a fuller understanding of the major political events of our time and potentially those yet to come.

HILLARY CLINTON WAS GREAT AT getting people to vote for
her. In fact, she was better at it than Donald Trump by nearly
three million votes. Unfortunately, the Trump campaign,
the media, the Democratic Party establishment, the Russian government, and even Clinton herself proved to have a
knack for getting people to *not* vote for her. There is no single
reason why Clinton failed to win the US presidency in 2016.
I'm certainly not going to claim it was solely because of spite.
However, when the question "What happened?" is asked, we
can't ignore the potential impact of spiteful politics.

Looking at American politics from the outside, the 2016
presidential election seemed to be a punishment election.
One could be forgiven for thinking that voters went into the
voting booth thinking not "whom should I support?" but
"whom should I hurt?" Such punishment was often aimed
at Hillary Clinton, and it could be relatively cheap. A potential Hillary voter might simply stay home on election day.
Punishment could also bear a moderate cost—for example,
taking the time and energy to go to the polls to vote for a
left alternative, perhaps the Green Party's Jill Stein. Yet punishment was also possible at greater cost. A potential Hillary voter could cast their ballot for Donald Trump in the
knowledge that his presidency could hurt them, their country, and the world. All three forms of punishment seemed to
have happened. As a result, 2016 saw the spiting of Hillary
Clinton.

We may doubt that people would vote against their best interests. Yet, as we have seen, a significant minority are prepared to spite. To establish that people went so far as to vote for Trump to spite Clinton, or at least that spite was a contributing factor in their decision, we would need to show two things: first, that people were motivated to punish Clinton, and second, that they did so at a personal cost by voting for Trump knowing he could harm their interests. There is evidence for both.

First, take the results of a CNN exit poll that asked people their opinion of the presidential candidate they voted for.[2] A quarter of respondents said it wasn't that they strongly favored the candidate they had voted for, or even that they simply had reservations about that candidate. Instead, they voted for their candidate because they disliked the opponent. Of the people who felt this way, 50 percent had voted for Trump and 39 percent had voted for Clinton. People were hence more likely to vote for Trump because they disliked Clinton than they were to vote for Clinton because they disliked Trump.

Polling by the Pew Research Center found similar results.[3] Their research discovered that 53 percent of Trump voters said they were voting against Clinton more than they were voting for Trump. This was a dramatic change from previous US presidential elections. In both 2008 (Obama vs. McCain) and 2000 (Bush vs. Gore), around 60 percent of each candidate's supporters said their vote was mostly for their candidate. Clearly, 2016 was a punishment election, but was it a costly punishment election?

It appears to have been, because, turning to our second criterion, evidence shows that some people voted for Trump knowing it would have a cost for them.[4] For example, you would not expect the pool of people who said they would feel negative if Trump won to include anyone who voted for Trump. Yet 13 percent of this pool turned out to be Trump voters. Similarly, you would not expect the group of people who said they would be concerned if Trump won to be Trump voters. Yet a full third of these people emerged as Trump voters! Finally, one in three voters said they would be scared if Trump won. Of these, 2 percent were Trump voters.

These data show that people voted for Trump despite conceding that they would feel negatively, concerned, or straight-out frightened if he won. Fewer people (between a third and a half less) voted for Clinton with the same reservations.[5] It is hence plausible that people voted for Trump because they disliked and wanted to punish Clinton, in the full knowledge that doing so could hurt them. They would have been voting for Trump to spite Clinton.

THERE ARE, OF COURSE, OTHER reasons why people may have voted for Trump—even while expecting trouble if he won—other than to spite Clinton. We can see this in voters' statements. As one Trump voter put it:

> A dark side of me wants to see what happens if Trump is in. . . . There is going to be some kind of change and even if it's like a Nazi-type change. People are so drama-filled. They want to see stuff like that happen. It's like reality TV. You don't want to just see everybody be happy with each other. You want to see someone fighting somebody.[6]

Similarly, a Bernie Sanders supporter stated, "Honestly, I want to vote for Trump—not because I agree with anything he says but because I'd rather have it all burn down to the ground and start over again."[7]

We may not agree with this sentiment (indeed, it is hard to understand how anyone with even the slightest familiarity with history could get close to supporting a "Nazi-type change"), but we may recognize it. Alfred, Bruce Wayne's butler, famously expressed this idea in the 2008 Batman film, *The Dark Knight*. As he put it: "Some men just want to watch the world burn."

The desire to burn the world to the ground may seem like spite. Indeed, it may well be harmful to both oneself and others in the short term. Yet, in the long term, it could be in some people's self-interest, as suggested by recent work that has studied the "need for chaos." This research, led by the Danish psychologist Michael Bang Petersen, began by examining what led people to spread political rumors online.[8] Petersen concluded that people weren't simply doing it to boost their own party or to hurt "the other guy." He proposed that people did it because they were pissed off with the state of society and their place within it.

To test this idea, Petersen and colleagues created a "Need for Chaos" questionnaire, which included items such as "I get a kick when natural disasters strike in foreign countries"; "I fantasize about a natural disaster wiping out most of humanity such that a small group of people can start all over"; and "I think society should be burned to the ground." They then studied how Americans, Danes, and non-Western immigrants responded.

The first thing to note was the number of people who endorsed such extreme statements. Ten percent agreed that society should be burned to the ground. Twenty percent agreed that problems with societal institutions cannot be fixed and so we should tear them down and start over. Those who displayed a greater need for chaos were more likely to share hostile political rumors online and to have a violent activist mindset.

Obviously, agreeing with such extreme statements on a questionnaire is very different from performing those actions. The study didn't start with people who had committed such acts and then work back to discover that their motivation was a need for chaos.[9] Nevertheless, those with such a need are potentially people who, were they to get hold of one of Nick Bostrom's black balls, could imperil us all.

So what is happening here? Petersen argues that a need for chaos reflects a wish for a clean slate or a new beginning. Individuals who feel this way are likely to be those who would benefit from the collapse of the status quo, people who seek status but lack it. The American Founding Father Alexander Hamilton captured this sentiment when he identified a potential threat from a perverted class of men who hoped to aggrandize themselves by the confusion of their country.[10]

Petersen and colleagues propose that chaos incitement is a strategy of last resort used by marginalized status seekers. They reported evidence consistent with this hypothesis. Having a high need for chaos was associated with being young, less educated, and male. It was also associated with higher levels of loneliness and the feeling that one was positioned on a lowly rung of the social ladder.

Peterson's study suggests that social marginalization would be most likely to create a need for chaos in people who had the skill set to navigate antisocial situations. This includes a lack of empathy and (in males) physical strength. Furthermore, Peterson argues that rising inequality and dissatisfaction with both democracy and quality of life tap into the processes that catalyze a need for chaos. They simultaneously activate and frustrate status striving. To defuse this desire we need to ensure that everyone has dignity, a stake in society, and, as we will see in the next chapter, a prosocial sacred value to struggle for.

Another, although less well supported, hypothesis for why people voted for Trump despite expecting trouble comes from the theory of "elite betrayal." This was proposed by Rafael Di Tella and Julio Rotemberg of Harvard Business School.[11] Their idea starts from the observation that, given a choice, people prefer to be screwed due to bad luck rather than because someone else took advantage of them.[12] If you hold that attitude and are concerned about government corruption, you are more likely to vote for an incompetent politician than a competent one. The incompetent candidate may blunder and accidentally make you worse off, but the competent one could intentionally screw you.

Researchers investigated this idea one week before the 2016 US presidential election. They tested whether emphasizing the importance of competence in politicians altered the way people were likely to vote. An effect was found in a group of voters that seemed to be open to populism: nonrural white voters with low levels of education. Sixty-three percent of this cohort thought Clinton was more competent than Trump. One might expect that, when given information on how important competence is in politicians, this group would be more likely to vote for Clinton. But no, the group now became 7 percent more likely to vote for Trump. Before concluding that this means they were worried Clinton would betray them, we should note that the statistics the authors report suggest it could be a chance finding.[13] More convincing evidence of their theory is needed.

━━━

IF SOME PEOPLE WERE VOTING to punish Hillary, what were they punishing her for? If there was one thing Hillary Clinton did not want to do, it was to make liberals, who would typically vote for her, question her fairness. This is because, when making moral decisions, liberals place more emphasis on fairness than conservatives. We know this thanks to the work of the American psychologist Jonathan Haidt.

Haidt's "moral foundations theory" argues that people's moral concerns revolve around six dimensions: care/harm, fairness/cheating, loyalty/betrayal, authority/subversion, purity/degradation, and liberty/oppression. His work has found that people of different political affiliations place different importance on these dimensions. Of particular relevance here is that liberals place more weight on the fairness/cheating dimension than conservatives.[14] Reflecting this, Hillary's campaign stressed many issues related to fairness: women's rights, campaign finance reform, and income inequality.

Given the importance of fairness to liberals, Hillary needed to be seen by Democratic Party members to have won the party nomination fairly. Any such violations were likely to trigger rage. Unfortunately, she did not achieve this aim. Her battle with Bernie Sanders in the Democratic primaries led some to perceive that she had gained the nomination unfairly. Many Sanders supporters believed that the Democratic Party establishment was not impartial, as they felt it should have been. Instead, the Sandernistas saw the party establishment as being on the side of Clinton. They had plenty of reasons to feel this way.

First, large numbers of Sanders's young supporters weren't registered to any political party. In some states, this meant they were barred from voting for him in the primaries. Second, there were the superdelegates, party officials who had an outsized voice in choosing whom the party put forward as its nominee for president. They had the potential to override the votes of average party members. One member of Sanders's campaign recalls the Democratic National Convention:

> I'll never forget watching the primary votes being counted for Michigan, one of the key states that decided the 2016 election. Sanders's pledged delegate count—which reflected the number of votes he received from rank-and-file Democrats—exceeded Clinton's by four. But after the superdelegates cast their ballots, the roll call registered "Clinton 76, Sanders 67."[15]

Similarly, Sanders received 60 percent of the popular vote in New Hampshire, yet, thanks to superdelegates, New Hampshire's votes were split evenly between Sanders and Clinton at the convention.[16] Sanders himself complained that this "rigged system" blocked his path to the nomination.

Whether true or not, the effect of his statement played straight into the narrative arc of Donald Trump, which we will come to momentarily.

WikiLeaks raised more fairness concerns by releasing emails, given to them by Russian intelligence, showing that senior Democratic National Committee members were keen to undermine the Sanders campaign.[17] Methods broached included targeting his religious faith.[18] The Democratic National Committee apologized, but the donkey had already bolted.

Sanders himself was also instrumental in creating perceptions that Clinton did not act fairly. While he downplayed both the release of the Democratic National Committee emails and Clinton's use of a private email server, he portrayed Clinton as a tool of the establishment who cozied up to Wall Street elites. This fed into a generational issue. For many older women, Hillary was a pioneering feminist. Yet for some younger women, she had come to represent the dominant class.[19] For them, Hillary was a symbol of the establishment, a white, pantsuit-wearing insider, protecting her class interests. Sanders highlighted this perception, which arguably had a devastating effect in the presidential election itself.

Whether or not the 2016 Democratic primaries were rigged in Clinton's favor,[20] 27 percent of Democratic voters thought they were. Only half thought she had won the nomination in a fair contest. For liberal voters to question Clinton's fairness was potentially disastrous.

Once Clinton had won the Democratic nomination, Sanders needed to undo the perception of unfairness. He was well aware of the possibility of spite voting. "This is not the time," he warned before the election, "for a protest vote."[21] Sanders did try to reverse such views, including those he had contributed to. "She won fair and square, right?" Wolf Blitzer asked him on CNN.[22] "Yes," said Bernie. Yet when Blitzer followed up with the question "So it's over?," Sanders's answer was not unambiguous: "Well, what's—no. What's over is, you go into the Democratic Convention. But, you know, what is over is the fact that she has more delegate [*sic*] than I do. I'm not arguing with any of those." It is unclear what his supporters took from this comment.

The obvious way to try to get Sanders's supporters back in the fold was to appeal to their party loyalty. Of course, many were not in the party. But even for those who were, given that the moral foundations of liberals are more solidly grounded in fairness than in loyalty, such a call struggled to compete with claims of unfairness.

A 2017 study exposed liberals to two sets of opposition messages to see which would make them least supportive of Hillary.[23] One set of messages was based in values related to fairness—for example: "Clinton is willing to sacrifice fairness and equality to achieve her own goals," positioned next to a picture of Clinton beside a Wall Street sign. The other set of messages was grounded in loyalty and its opposite, betrayal—for example: "She failed our ambassador and soldiers in Benghazi," with a picture of Clinton next to an open envelope with an email symbol inside. The researchers found that liberals were less likely to vote for Clinton if exposed to the unfairness argument than if presented with the betrayal argument.

Once Clinton won the Democratic Party nomination, there was also the job of getting Democrats who disliked her to vote for her. Some campaigners took an approach that respected the autonomy of voters. "My request to those of you who still don't want to vote for Hillary Clinton: Please reconsider," wrote the former secretary of labor Robert Reich.[24]

Others took a stronger approach. Person after person stood up to tell voters not only that they should vote for Clinton but that they *must* vote for her, that they *had to* vote for her, that there was no other choice. Sometimes this was stated directly. "You *must* vote for Hillary Clinton," demanded the singer Katy Perry.[25] Sometimes it was stated indirectly. "There's a special place in hell," opined Madeleine Albright, former secretary of state under Bill Clinton, "for women who don't help other women."[26] Sanders supporters took this as Albright telling them how to vote. Such sentiments could not have been better designed to trigger the Braveheart Effect. Although some voters, including Sanders supporters, agreed with the view that they had to vote for Clinton,[27] it is entirely plausible that these tactics pushed others into spiteful courses of action.

Indeed, for many Sanders supporters, righteous anger surrounding the nomination process was still bubbling away when the presidential election came around later that year. This anger had the potential to trigger spite in the form of punishment, at a variety of costs. Some stayed home rather than vote for Clinton, which in certain cases probably represented an act of punishment. Sanders supporters who did go to the polls mainly voted for Clinton, even if with a heavy heart. Yet a significant minority voted for Trump. To be precise, 12 percent of people who voted for Sanders in the 2016 Democratic presidential primaries voted for Trump in the presidential election. If 12 percent of voters in the swing states of Michigan, Wisconsin, and Pennsylvania had voted for Clinton instead, or even just stayed home on election day, she would have won those three states and the White House.[28]

There is nothing particularly unusual about 12 percent of a primary candidate's supporters eventually voting for the other party. A similar percentage of Republican primary voters ended up voting for Clinton. However, we might wonder if the motivation of Trump-voting Sanders supporters was more rooted in spite than that of the average switch voter. Clearly, some supported Trump because they shared certain views with him.[29] But when it came to the issues, there were remarkably few points of overlap between the two candidates. The anger whipped up during the primary is likely to have pushed a percentage of Sanders supporters into voting for Trump to spite Clinton.

Two days before the presidential election, the magazine *Marie Claire* ran an article titled "How to Calmly Talk to Your Friend/Aunt/Brother-in-Law About Why Voting for Trump Just to Spite Hillary Is a Really, Really Bad Idea." Online comments during the election reflected a spiteful sentiment. "I will not only not vote for Hillary Clinton, if she becomes the nominee," double-negatived one online commentator; "I'll vote for Trump out of spite to make the Hillbots suffer for refusing to get behind the only actual liberal in the race."[30]

Such lingering perceptions of unfairness stemming from the Democratic primaries may not have been enough on their own to bear much impact on the election. However, both the Trump campaign and the mass media amplified this narrative.

———

A CORE PART OF DONALD Trump's message was that Hillary Clinton had acted unfairly and that voters should punish her. His moniker "Crooked Hillary" encapsulated this view perfectly. According to Trump she was "the most corrupt person to ever seek the presidency."[31] Such statements formed part of Trump's strategy of cultivating the idea that ordinary people faced an unfair political system created by insider elites who wrote rules to "keep themselves in power and in the money."[32] The billionaire in the baseball cap positioned Clinton as one of this elite. Trump argued that Clinton acted unjustly by protecting the interests of Wall Street over those of the average person. He claimed she was "owned" by the financial class, whereas he, being a billionaire, wasn't in their pocket. His narrative also helped mitigate perceptions of his problematic behavior. Trump may have gotten away with violating standards of decency without being punished because some voters took his outrageous statements as evidence of his willingness to stand up to the elites.[33]

Trump also picked up and ran with Sanders's claims of unfairness. "To all of those Bernie Sanders voters who have been left out in the cold by a rigged system of superdelegates," Trump bellowed, "we welcome you with open arms."[34] Trump's message that Clinton was a tool of Wall Street echoed Sanders's claims that she wouldn't stand up to the financial elites, Big Pharma, and the insurance and fossil fuel industries. Trump poured gasoline on a fire that Sanders had started.

Trump also cited other examples of Clinton's alleged unfair behavior. He claimed she had accepted donations to the Clinton Foundation in exchange for access to herself and her State Department. The roots of this accusation lay in Russian-facilitated WikiLeaks revelations.[35] Trump argued that justice was circumvented because the use of a private email server while working as secretary of state would have led to anyone else being arrested. When the FBI recommended that Clinton not be charged over her use of a private email server, Trump took to Twitter to cry "Very very unfair!"[36] Again and again, Trump pushed the argument that Clinton didn't play by the rules.

Perhaps the most absurd of all the spite-inducing claims made about Hillary Clinton was that she wasn't qualified to be president of the United States. Trump's campaign promulgated the idea that she had ridden her husband's coattails into positions of political power. Another allegation was that she was being considered for the highest office in the land only because the politically correct had decided it was time for a female president. In April 2016, during the same interview in which Trump introduced the world to the term "Crooked Hillary," he also claimed, "The only thing she's got is the woman card." The implication was that Clinton was unqualified for the job and that her claim to it lay solely on a gender-based affirmative action principle.[37]

This was ludicrous. As President Barack Obama put it, "I can say with confidence there has never been a man or a woman—not me, not Bill, nobody—more qualified than

Hillary Clinton to serve as president of the United States of America."[38] We saw earlier how perceiving others as enjoying unmerited gains can trigger a spiteful response. The narrative that Clinton stood undeservingly on the threshold of the presidency had the potential to be a powerful stoker of spite.

━━━

IF THE ONE THING THAT Hillary did not want to do with liberals was to appear unfair, the one thing she did not want to do with potential Trump voters was to appear to think she was superior to them. People's belief that you have more than they do, whether it be more money, morals, or social status, can trigger counterdominant spite.

In arguing that Clinton thought she was superior to the average person on the street, Trump supporters had a vat of material to draw on. Indeed, before the 2016 "Crooked Hillary" attack, prior criticisms had often centered on Clinton's "ostentatious virtue and moral superiority."[39] One such alleged example was Clinton's infamous cookie comments. In 1992 she reportedly said, "I suppose I could have stayed home and baked cookies and had teas, but what I decided to do was fulfill my profession." The way this was reported encouraged women to think that Clinton viewed herself as better than them. As one voter put it, "I was ready to like her. Not now . . . after what she said . . . she obviously doesn't have respect for what I do."[40] This quotation would repeatedly resurface, including in the 2016 election cycle. Although Beyoncé used part of the quotation to encourage people to vote for Clinton,[41] it was more often used by others to portray Clinton as arrogant. In reality, Clinton had gone on to say that by fulfilling her profession she wanted to give women the right to choose what they wanted to do. The media did not report this. They wanted outrage. They got it.

Another example, also from 1992, was when Clinton said, after being questioned about her husband's infidelity, that she "wasn't sitting here, some little woman standing by her man like Tammy Wynette." Clinton had been referring to the Tammy Wynette song rather than the performer herself. Nevertheless, Wynette took it personally and responded in the press, saying, "I can assure you, in spite of your education, you will find me just as bright as yourself."[42] Clinton's response to hearing this, an eye roll and a head slap, did not improve matters.

The Trump campaign also had contemporary material to draw on to denigrate Clinton as morally condescending and out of touch. Her leaked speeches to Goldman Sachs showed her saying things to bankers such as, "You are the smartest people."[43] Yet by far the most egregious example was Clinton's potentially fatal "basket of deplorables" comment. During a speech at a campaign fundraising event in September 2016, Clinton stated, "You could put half of Trump's supporters into what I call the basket of deplorables. . . . They're racist, sexist, homophobic, xenophobic—Islamophobic—you name it."

Now, let's be honest about two things. First, we can argue about the size of the basket, but we cannot argue that there was no basket. In her 2017 book, *What Happened*, Clinton defended her comments with statistics. These included the results of a 2016 survey that showed white Republican voters held more racist views than white Democratic voters.[44] Later research also supported concerns about the association of these "-isms" with *some* Trump voters, as I will discuss in Chapter 7. Second, we all know that some truths cannot be spoken during election campaigns. It is hard to quantify the impact of Clinton's "deplorables" comment. Jonathan Haidt suggests it "could well have changed the course of human history."[45] It does, at least, seem likely to have encouraged a spiteful backlash.

Another force the media and the Trump campaign summoned against Clinton was spite in the form of do-gooder derogation. Even when she had been First Lady of the United States, the press had mockingly referred to Clinton as "Saint Hillary."[46] In 2016, her identity politics and rights-based campaign had the potential to make people think she was better than them, which in turn had the potential to trigger a spiteful reaction.

This raises the question of how one can campaign for and do the right thing while minimizing do-gooder derogation. It is possible that one can't. One may have to be willing to be a martyr for one's message. If you let people shoot the messenger, they feel less urge to also burn the message.[47]

This intriguing idea, that we tolerate challenging messages but not their messengers, comes from a study of vegetarianism. Members of one group of participants were asked to reflect on how vegetarians would see them, exposing them to the potential threat of feeling judged. They were then asked to rate how kind, dirty, and stupid they thought vegetarians were. The members of another group were asked to do the rating of vegetarians first and then were asked to reflect on how vegetarians would see them. At the end of the study, both groups were asked about their views regarding eating meat.

As expected, the group that was exposed to the threat of feeling judged by vegetarians before rating them disliked vegetarians more. However, surprisingly, they then went on to be *less* supportive of eating meat than the group who answered the questions the other way around. This suggests that letting people negatively judge a do-gooder increases the chance they will accept their message. Shooting, crucifying, or simply failing to elect the messenger may encourage the acceptance of their message. Not all messengers will be consoled by this.

By the end of the 2016 US presidential campaign, a wide range of actors, including Donald Trump, Bernie Sanders, WikiLeaks, and Vladimir Putin, had worked to portray Hillary Clinton as having violated norms of fairness. Trump's campaign was also able to depict Clinton as believing she was morally and socially superior to the average voter. Add voters being told they had no choice but to vote for Clinton, and the stage was set for voters to act spitefully against her.

━━━

THE FORCES ARRAYED AGAINST HILLARY Clinton are not uniquely American. Around the world, populists frame their message as supporting the interests of ordinary citizens against elites.[48] The motivation of most voters who support populist candidates is self-interest. They believe that the populist will make their lives better. But as we have seen, there is a small and potentially influential tranche of voters who are primarily interested in spiting elites. Such individuals will either value harm to the elites over their own financial interests or will find concepts of "taking back control" more important than immediate economic gains. Molly Crockett argues that "what we're seeing all around the world with the rise of populism is a kind of global large-scale, society-level version of an Ultimatum Game."[49] As she notes, Brexit is another example of the phenomenon.

In the United Kingdom's 2016 referendum on its membership in the European Union, 52 percent of its citizens voted for the UK to leave. Many European politicians view this as a lose-lose outcome for the UK and for Europe,[50] raising the possibility that at least part of the vote to leave was spiteful. During campaigning, politicians seemed to sense that spite would rear its head. Both the Conservative prime minister, David Cameron, and the leader of the Scottish National Party, Nicola Sturgeon, voiced concerns that people would vote to leave as a protest against them. Sturgeon explicitly urged voters, "Don't cut off your nose to spite your face. Don't base your decision about the future on a gripe or a grievance you might have." Cameron's worries may have stemmed from his own "deplorables" moment. He had called a segment of Brexit supporters "fruitcakes," "loonies," and "closet racists."[51] Although these comments had been made a decade earlier, they had not been forgotten.

Some voters' grievances were not only aimed at domestic politicians. The British media had a long history of stirring up dislike against European politicians. "Up Yours, Delors," said an infamous headline of a British tabloid in 1990. This, and the shameful diatribe that followed, expressed dissatisfaction at proposals, alleged to threaten UK autonomy, by then European commission president Jacques Delors.

The idea of voting to hurt a perceived elite of UK and European politicians was undoubtedly in the air. The campaign for the UK to remain in Europe tried to ward off this notion by appealing to voters' financial self-interest. The Remain campaign relentlessly pushed the message that to leave would be to suffer a tremendous economic cost. David Cameron claimed that leaving Europe would "put a bomb under our economy."[52] The chancellor of the exchequer, George Osborne, repeatedly cautioned that to leave Europe would result in job losses and tax increases. The governor of the Bank of England, Mark Carney, warned of a possible "technical recession" if the UK left Europe. "Stronger, Safer and Better Off" ran the official Remain campaign slogan. All this emphasized to voters that voting to leave to harm the elite would be a *costly* punishment. Yet, as we have seen, people are quite willing to embrace such spiteful actions.

The campaign to leave pushed every possible button to generate spiteful sentiment toward the campaign to remain. First, it amplified a narrative that Europe was acting unfairly toward the UK. Europe wasn't allowing the UK to make its own laws. Instead, "unelected bureaucrats" in Brussels were deciding the UK's fate. Europe wasn't allowing the UK to control its borders. The British people hence faced an influx of migrants who would rape them, take their jobs, and destroy their health service.[53] As we have seen, if people feel that a norm of fairness has been violated, counterdominant spite will push them to make the offender pay, even if such spite comes at a personal cost. Naturally, the Leave campaign stressed the self-interest motive too. The Leave campaign's slogan was "Let's Take Back Control." The United Kingdom Independence Party (UKIP) stressed, "We Want Our Country Back." This emphasized that, even if Brexit had an economic cost, the freedom dividend would offset it. Indeed, by making people's freedom and autonomy feel threatened, the Leave campaign was able to trigger the Braveheart Effect.

The Leave campaign was not merely able to portray the European project as unfair in order to trigger counterdominant spite. It was able to represent Europe as violating a sacred value: that of a gloriously independent United Kingdom. This was the country of Churchill. A country that had fought on the beaches and the landing grounds. A country that had fought in the fields and in the streets. A country that would never surrender. Activation of a "sacred value," as we will see in Chapter 7, took the choice to leave Europe out of the arena of cost-benefit analysis. It became one's duty to leave, consequences be damned. And as everyone knows, England expects that everyone will do their duty.

The Leave campaign also portrayed supporters of the Remain campaign as thinking they were superior to the average voter, which helped to trigger counterdominant spite in voters. The Leave campaigner Nigel Farage repeatedly referred to a Westminster elite that looked down on the people. Many Remain supporters did not help dispel this perception. The chairman of the Labour Party's campaign to remain in Europe stated, "We are the reasonable people, the Leave side, they are the extremists," adding that the Leave side showed a "certain kind of mentality that is not rational."[54] Remainers made people feel like idiots for considering voting to leave.[55] It was the basket of deplorables all over again.

Indeed, after the campaign, Farage would claim that Remainers had "given the impression that those who voted for Brexit didn't know what they voted for, we're thick, we're stupid, we're ignorant, we're racist. It's given people a sense that those who voted Remain feel morally superior to those who voted Leave. . . . They genuinely think because they are superior to the Leavers."[56] Voting Leave to spite them would bring them down a peg or two. Voting Leave would also stick it to the experts who said that it was best to remain. Recall from earlier that it was during the referendum campaign when Michael Gove came out with his doozy that "people of this country have had enough of experts."[57] This was the best expression of our times of counterdominant spite.

When voters stepped into the privacy of the voting booth, the Leave campaign had set the scene for spite. Some voters felt they faced the following dilemma: stay in Europe and get some economic benefits, but violate a sacred British value, see the smug elite celebrate, and lose your autonomy. When remaining was framed like this, it is not surprising that a percentage of people spitefully rejected it.

But maybe no one voted Leave in the expectation of the economy getting worse, and therefore no one really acted spitefully. There are two replies to this objection. One comes from data, the other from psychology. In terms of the data, in the days before the referendum, polling showed a small group of people intending to vote Leave who believed that Britain would be economically worse off if it left the

European Union.[58] Four percent of people planning to vote Leave felt this way. Had they voted Remain instead, the UK would still be in Europe. Psychology suggests that more than this 4 percent might have initially thought that Brexit would be costly to the UK economy. We change our beliefs to fit with our actions. Upon deciding to leave, some voters may have changed their minds about the economic impact to make their new stance seem more justifiable.

This 4 percent, although a small group, were nevertheless important to the outcome of the referendum. They also show us something else important about spite. Most likely, this group of voters valued the need to "Take Back Control" over the economic losses they would suffer. When European politicians describe Brexit as lose-lose, they refer to economic outcomes. They underestimate the value of voters feeling free and the economic price they are prepared to pay for this. They also underestimate the pleasure of counterdominant emotions such as the joy of perceived justice and schadenfreude. So, did the noneconomic benefits gained from voting Leave offset the economic costs for this 4 percent of voters? Did they net out into an overall gain for those voters, making their ballots selfish rather than spiteful? Is this even a question we can answer?

We could argue that voting Leave, in the knowledge that there would be economic losses, was at least a risky move, making such an action weakly spiteful. We could also argue that the benefits of freedom that people anticipated gaining, after escaping from the perceived yoke of Europe, genuinely did offset any economic costs, making their votes selfish, not spiteful. Contrastingly, we could argue that if the only compensation for economic losses was emotional, in the form of schadenfreude, then this made the act still costly and hence spiteful. My view, from the discussion in the paragraphs above, is that at least some Leave voting could be described as spiteful. Yet, ultimately, we are brought back to the question from the Introduction: who determines if an act was costly and therefore spiteful? The choice of those who voted to leave are in danger of being labeled spiteful by people who disagree with this voting pattern and are looking to denigrate it. The term "spite" is in danger of being used to *do* something rather than to *describe* something. Everything is political, even the question of what is spiteful.

━━━

A QUESTION WORTH PONDERING IS whether spite voters are likely to be more prevalent today than in other periods of history. One way to approach the query is to use the "structural-demographic theory" of the Russian American scientist Peter Turchin, who argues that societies rise and crash in cycles of about two hundred years.[59] Turchin demonstrates this pattern for a wide range of societies, from France to China to the United States. He argues that in the late 1970s, the United States entered a period of decline that he calls an "age of discord." He suggests it resulted from three factors.

First, there was the influence of the mass of the people. Due to factors such as immigration, the entrance of women into the workplace, and the outsourcing of manufacturing to other countries, the supply of labor in the United States exceeded the demand for it. This forced wages and living standards down. The real wage of the American male today is less than it was forty years ago, which has led to intense dissatisfaction. Such a state of affairs is necessary for societal catastrophe, but not sufficient. An additional factor is needed: what Turchin identifies as elite overproduction.

Elite overproduction occurs when employers and those in positions of power start to make more and more money as a result of declining wages, creating an ever-increasing number of people who want to rise into the elite. But there are only a finite number of elite positions: president, members of Congress, Supreme Court justices, CEOs, university presidents, chief surgeons. As a result, elites become infected with rivalry and factionalism.

A new counterelite can emerge. Their path to power is through mobilizing the masses against the ruling elite in an attempt to grab the reins of power for themselves. The victory of Trump was not just due to the dissatisfaction of white males in America's Rust Belt. Any uprising of this group would have failed if faced by a unified elite. In 2016 Trump ousted the Democratic elite by mobilizing the masses against them. It was not the "deplorables" who caused the electoral upset, but the "frustrated elite aspirants." Such situations, Turchin argues, can culminate in political violence, civil war, and revolution. The best way to address the underlying problem is to correct the oversupply of labor. People need jobs. Competition may produce great breakthroughs and create efficiencies, but we must be aware of its potential for spite and revolt.

BEFORE MOVING ON, LET'S EDGE out a bit farther along the branch of the known and consider a more speculative political implication of the work we have discussed. Earlier, I pointed out how increases in serotonin are associated with a reduced willingness to pay to punish unfair acts. Given the ubiquity of medicinal drugs in our society that alter serotonin levels, it is worth reflecting on what the political impact of those drugs could be.

Let's start with the obvious one: the selective serotonin reuptake inhibitor (SSRI) class of antidepressants—drugs such as Prozac, Zoloft, and Celexa. By definition, these medications increase people's serotonin levels. They offer a substantial benefit for people with severe depression. Yet the benefits of antidepressants over placebo are minimal or nonexistent, on average, for patients with mild or moderate symptoms.[60] Some people hence appear to be using them with no clinical benefit while still potentially suffering their side effects,[61] which can include sexual difficulties, emotional numbness, and suicidality. But could another side effect be a reduced willingness to pay a personal price to protest unfairness?

If so, then this effect could be widespread enough to have a significant societal impact. Between 2011 and 2014, approximately 13 percent of people age twelve and older in the United States reported taking antidepressants in the past month.[62] In England, a country of fifty-six million people, seventy-one million prescriptions for antidepressants were written in 2018.[63] Are antidepressants, by increasing serotonin levels, making the population less willing to spite and more willing to go along with unfairness? What is the full cost of these drugs?

A class of drug widely used among women that affects serotonin levels relates to contraception. In the United States, between 2015 and 2017, 8 percent of women were taking oral contraceptives, and another 7 percent were using long-acting reversible contraceptives, including contraceptive implants and intrauterine devices.[64] Determining in which direction such contraceptives push serotonin is complicated. The combined pill contains both estrogen and progestin, which have opposite effects on serotonin levels, increasing and decreasing it respectively.[65] Yet both the mini pill and intrauterine contraceptive devices (such as the Mirena coil) release progestin only. This will decrease serotonin levels and could, in theory, increase women's anger at unfairness. The potentially liberating effects of contraception may be greater than we know.

Although I have finished this chapter on a speculative note, spite clearly has a significant effect on the political world. Let's now turn to another area of life that spite can shed light on: the sacred.

Seven
SPITE AND THE SACRED

St. Paul counseled the Romans to "avenge not your-selves . . . for it is written, Vengeance is mine; I will re-pay, saith the Lord."[1] Two thousand years later, a similar sentiment was voiced. In the film *Pulp Fiction*, Samuel L. Jackson's character, Jules, drew on the words of the prophet Ezekiel: "And you will know my name is the Lord when I lay my vengeance upon thee."[2] Although differing in their fidelity to the message, the words cited by Paul and Jules suggest that vengeance is something to be delivered from above. Why did this idea arise? Could it be that humans, reticent to spite others due to the potential for retaliation and the lack of esteem for second-party costly punishment, outsourced it to the divine? Are gods a cheap form of spite?

Most humans believe in a god. Christians and Muslims alone make up 55 percent of the world's population.[3] Conceptions of god vary across the world. Yet there is a widespread belief in a being who knows what we have done, who knows right from wrong, and who will punish us for transgressions.[4]

At a certain time in our history, such a being had the potential to come in particularly useful. When humans began to live in agricultural societies, that meant they lived in much larger groups than they had as hunter-gatherers. The cost of punishing others rose significantly. Individuals in agricultural societies could amass more wealth and power than any hunter-gatherer could have done. They could hence retaliate against punishment with great force. The mechanisms that had previously promoted cooperation in small groups began to break down under this stress. People needed a new way to promote cooperation in large groups. Genetic evolution could not provide a quick answer, so humans turned to culture. They needed to create an authority to deliver punishment. This could have been an earthly institution, but it could also have been an otherworldly one: a god.[5]

The psychologist Kristin Laurin argues that because punishing others for unfair behavior can be very costly to us, we invented gods to do it instead. Laurin notes that gods are "most likely to emerge in large societies, or those with resources shortages, both of which have particularly high needs for regulating and enforcing cooperation."[6] But how do you get the powerful to believe in a deity that dislikes them?

As Laurin and colleagues point out, the major religions have ways of making you believe. Religions play on biases in human psychology, such as our tendency to detect an intelligence behind events caused by chance. They get people to perform costly rituals that signal to others that they truly believe.

Evidence supports the idea that gods functioned to provide low-cost punishment that promoted cooperation in large societies.[7] The more a society believes in the existence of a punishing god, the more their population complies with social norms. People can fear god more than they fear their peers. The more you believe in this god, and in concepts such as heaven and hell, the less likely you are to violate social norms. The more you believe that god punishes, the less you will cheat on tests and the less you will engage in costly punishment yourself. Yet to see these effects, people need to be reminded of their religious beliefs before they act. In religious societies, such reminders are everywhere.

In addition to being a low-cost way to deliver counterdominant spite, religion may also be used for dominant spite. Nietzsche argued that Christianity was a dominance mechanism. Its morality lauded those lower down the social ladder. The meek are the blessed and the first shall be last. Nietzsche called this a "slave morality."[8] It is reminiscent of Richard Wrangham's phrase the tyranny of the underdogs, from Chapter 2. Yet Nietzsche did not view slave morality simply as a way to pull down the "great man." Rather, he saw it as a way to reverse the social hierarchy, to turn the tables.

It did not seek equality but rather new masters. The weak and the meek became the strong and the great.

———

THE IDEA THAT VENGEANCE IS the sole remit of a god is not something all believers subscribe to. The most salient example are those prepared to pay the ultimate price to harm others: religiously motivated suicide bombers. Although, as we have seen, there are many people prepared to spite others, suicide bombers are thankfully relatively rare. The costs to both oneself and others naturally deter most people.[9] Yet they do not deter everyone.

There have been approximately thirty-five hundred such attacks in the past thirty years.[10] Can we map a path from everyday spite to a suicide bomber? And, if so, what pushes those particular individuals over the edge? Before embarking on this line of inquiry, I should stress that to understand why someone acts the way they do is not to condone it. It is possible to accept that suicide bombers have their reasons for their actions while still disagreeing with their reasoning and utterly condemning their actions.

Suicide bombers are not simply the most spiteful people in the room. There is an ingredient that, if combined with spite, increases the chance of it being lethal. Surprisingly, that ingredient is altruism.[11] The difference between a terrorist and a common criminal is that the former are more likely to believe they are acting altruistically.[12] This has interesting implications. If suicide bombings turn out to be perverted "prosocial" acts, then we may wonder how terrorist organizations achieve such outcomes. Could this same knowledge be used to promote genuinely prosocial acts more generally? Could it help us save the world rather than blow it apart?

▬▬

SEVERAL ASSUMPTIONS NEED TO BE put to one side before we can explain the behavior of suicide bombers. They are not usually mentally ill. Neither are they already suicidal.[13] They are not typically from a broken background that makes them amenable to brainwashing, and they are not stupid. If anything, suicide bombers tend to be better educated and come from more privileged circumstances than the average person, although there does seem to have been some change in this demographic since the early 2000s.[14] Nevertheless, we are forced to look for reasons for such an individual's actions, rather than lazily attributing them to a broken brain.

If spite plays a role in suicide bombings, we need to know what type of spite is in play. Is it the counterdominant spite we saw in Chapter 2 or the dominant spite from Chapter 3? Suicide bombings appear unlikely to be primarily motivated by dominant spite. After all, it would seem difficult to gain a relative advantage over anyone if you are dead. That said, some suicide bombers are financially pressed people who have experienced shame and dishonor. They may see a suicide attack as a way to salvage their self-esteem, restore their personal and familial honor, and raise their social standing.[15] In many cases they do seem to be young men "ardent for some desperate glory," to use a phrase from Wilfred Owen.

It might seem to be an evolutionary glitch that our drive for status is so strong that we can forget we need to be alive to benefit from it. Yet inclusive fitness does not require us to be alive for our actions to benefit our genes. If our actions lead to a net benefit to our genes by helping relatives, they may be selected for by evolution. A 2008 study by the anthropologist Aaron Blackwell tested the theory of inclusive fitness in Palestinian suicide bombers.[16] To my knowledge, this was a non-peer-reviewed study and hence should be treated with caution. Nevertheless, its approach is interesting. Blackwell reported that male suicide bombers from middle-income families with many siblings (but not men from smaller, poorer, or richer families) could translate the payments received for suicide attacks into greater inclusive fitness. Specifically, the money their families received from Hamas or Hezbollah helped the suicide bomber's brothers pay the bride-price (money the groom's family pays to the bride's family), which is generally required in Palestinian society.

Blackwell noted that the demographics of Palestinian suicide bombers fit with the type of person his model showed could increase their inclusive fitness through such attacks. Such bombers tend not to come from poverty, are college-educated or employed, and have more siblings than the average Palestinian family. I am skeptical that evolution would be able to tune people's behavior so finely at this level. I would only expect to see such a pattern emerging after selection pressures had been acting on suicide bombers for many generations. Nevertheless, Blackwell's approach is a novel way to try to test a theory.

MOVING BEYOND INCLUSIVE FITNESS AND status-seeking considerations, what seems to be more likely to drive suicide bombings is counterdominant spite resulting from perceived fairness violations. People aren't generally recruited into terrorist organizations. They join.[17] Grievances bring them to the group. No amount of spiteful tendencies will induce terrorist activities unless there is a perceived wrong.

Such grievances typically relate to matters of unfairness. The perceived ill-treatment of Muslims in the Middle East by Western powers is a significant source of grievance.[18] This perceived ill-treatment is not limited to what is viewed as the military occupation of Middle Eastern lands; it is also viewed as being related to the myriad forms that humiliation can take.[19]

To illustrate the roots of terrorism in grievances, it is worth examining what the man behind the 9/11 terror attacks, Khalid Sheikh Mohammed (KSM), originally intended for that horrific day. As detailed in the *9/11 Commission Report*, KSM originally took a plan to bin Laden that involved the hijacking of *ten* planes.[20] Nine would be crashed into buildings, including the ones actually hit on 9/11, plus targets such as the FBI and CIA headquarters, the tallest building in California, and nuclear power plants. The tenth plane would be landed, with KSM on board, at a major US airport. KSM planned for the hijackers to kill all the male passengers and alert the media. He would then give a speech criticizing US support for Israel, the Philippines, and repressive governments in the Arab world. If this is the first time you are hearing this story, it is worth reflecting why.

Grievances can arise from personal mistreatment as well as, and sometimes in conjunction with, group mistreatment. The anthropologist and suicide terrorism expert Scott Atran relates what happened when he interviewed a young man who wanted to blow up the American Embassy in Paris. Atran asked him why he wanted to do it. The young man initially talked about the oppression of Muslims across the world. Atran pushed harder, asking, "Why did *you* want to do it?" Now the young man talked of how he had been walking down a street in Paris with his sister. She had accidentally bumped into an elderly Frenchman. He spat at her and called her a dirty Arab. "Then I knew," said the young man.[21]

To take another example, consider the female Islamist suicide bombers whom the media refer to as the Chechen Black Widows.[22] They appeared in the year 2000. Two Chechen women drove a truck full of explosives into the headquarters of a Russian Special Forces unit in Chechnya. Since then, most suicide attacks by Chechen rebels have involved women. The Black Widows are best known for their involvement in the 2002 Moscow theatre hostage crisis, in which more than a hundred people died. Images flashed around the world of nineteen women dressed in black with bombs strapped to them.

As the psychologists Anne Speckhard and Khapta Akhmedova explain, the roots of these women's actions lay in the gross injustices they had seen and suffered at the hands of Russia. Nearly all had lost close family members in operations carried out by Russian forces that ranged from bombings to "cleansings." Many had seen Russians abuse or kill one of their family members. They had lost sons, husbands, and brothers.

Some explanations deny that the sources of these women's actions were genuine grievances. There are claims that women were coerced to take part in terror activities by being kidnapped, raped, and drugged, claims that typically come from Russian journalists.[23] Yet such entry points are the exception rather than the rule. These women's deeds are motivated by perceived unfairness.

We have talked about how violations of fairness norms can trigger a spiteful response. Yet what is being violated in order to potentiate suicide is something even more powerful. What is being violated, Scott Atran argues, are sacred values.[24] These are nonnegotiable preferences. Examples of sacred values include the Palestinian refugees' right of return, ownership of Jerusalem, and Shariah law.

The crucial mark of sacred values is that their defense drives people to do things that go beyond the reasonable, regardless of risks or costs.[25] Rather than weighing the costs and benefits of a course of action, people committed to sacred values will simply do what they think is right.[26] Historical examples include the Spartans at Thermopylae, the defenders of the Alamo, Japanese kamikaze, and the 9/11 attackers. Sacred values are an exceptionally powerful way for a small movement to succeed, due to the motivating effects of such values, which promote spiteful actions.

When violence is in the cards and sacred values have been activated, people no longer think rationally but are driven by the power of moral emotions. A 2011 study by Jeremy Ginges and Scott Atran asked Israeli settlers in the West Bank about the dismantlement of their settlements as part of a peace agreement with Palestinians.[27] Settlers were asked about their willingness to engage in picketing and blocking streets. It depended on how successful they thought such protests would be. It was a rational choice. However, their willingness to engage in *violent* protest did not depend on how effective they thought it would be. Instead, it depended on how morally right they thought it was. Former US vice president Dick Cheney once said that terrorists "have no sense of morality." Yet Scott Atran argues that "you cannot want to harm or kill masses of people . . . unless you have a deep sense of moral virtue in what you do."[28]

Neuroscience provides evidence consistent with the idea that sacred values lead us to act without weighing the costs and benefits. When we think about sacred values, our brains behave very differently from how they respond when we think about nonsacred values. A 2019 study by Nafees Hamid and colleagues recruited thirty Pakistani men who supported Lashkar-e-Taiba, a militant group associated with Al Qaeda.[29] The researchers examined how the men's brains behaved in relation to the idea of sacrificing themselves for their values. When the men were thinking about their willingness to fight and die for sacred values relative to nonsacred ones, they exhibited reduced brain activity in their dorsolateral prefrontal cortices.

We met the dorsolateral prefrontal cortex in Chapter 2. We saw that it was involved in doing a cost-benefit analysis that allowed people to reject low offers in the Ultimatum Game. Reduced activity in this part of the brain when making decisions about being violent after the violation of sacred values suggests that cost-benefit analyses aren't being used. It is almost literally a "no-brainer" that sacred values should be defended. In contrast, this part of the brain was more active when nonsacred values were involved. The person now weighed the pros and cons of acting violently.

The researchers performed a follow-up study, in which they found that when the participants decided they weren't willing to fight and die for a value, their dorsolateral prefrontal cortex was talking strongly to another part of the brain called the ventromedial prefrontal cortex.[30] This part of the brain comes up with a valuation of an action, with "all things considered." In this case, it appears that the men's dorsolateral prefrontal cortices had done a cost-benefit analysis and decided that fighting to die for the nonsacred value wasn't worth it. It then passed this message on to the brain's valuation center. Yet the researchers found that when the men decided to sacrifice themselves for a sacred value, the dorsolateral prefrontal cortex was no longer talking to the ventromedial prefrontal cortex. In short, sacred values come with a "just do it" tag.

The obvious next question is how people can be made to perform cost-benefit valuations as to whether fighting and dying for a value is really a good idea. The researchers examined what happened when the men from Pakistan, while making their choices about whether to martyr themselves for a value, were told what their peers thought about the matter.[31] When they were told that their peers were less eager to fight and die, they were outraged. Nonetheless, their willingness to fight and die for both sacred and nonsacred values decreased. This was accompanied by an increase in activity in the cost-benefit analysis part of their brains, the dorsolateral prefrontal cortices. Their peers had made them think again.

An important implication of people's sacred values not being based in cost-benefit calculations is that you can't buy them off. In fact, trying to do so can make them even less likely to compromise or negotiate. A 2007 study examined how Palestinians and Israelis would react to a hypothetical peace deal that involved compromising over a sacred value— for example, Palestinians being asked to give up their sovereignty over East Jerusalem or their right of return.[32] It then examined how people would react to exactly the same deal, but with an added financial incentive. An example would be Israel paying Palestine $1 billion every year for a decade. The study found that people who felt the deal compromised a sacred value were more likely to oppose the deal if money was also offered. Adding the financial offer made them more outraged and more willing to violently oppose the agreement. This effect was not seen in those who didn't see the deal as violating a sacred value. As the Ultimatum Game suggests, suicide bombing is not something money can necessarily eliminate if perceived unfairness persists. You can't put a price on the sacred.

Nonsacred values can take on the mantle of the sacred. Social exclusion can make this happen. The neuroscientist Clara Pretus and her colleagues looked at the willingness of people to suffer for sacred and nonsacred values and how their brains reacted during such decision-making. They recruited thirty-eight young Moroccan men living in Barcelona who said they would engage in violent acts in defense of jihadist causes.[33] These men then played a computer game called Cyberball, in which a virtual ball is passed between players. The computer is programmed to rig the procedure to make a certain player feel either socially included (they get passed the ball a lot) or socially excluded (the ball is rarely passed to them). The designer of Cyberball came up with the idea after being excluded from a game of frisbee and being surprised at how bad he felt about it. Although Cyberball is a simple game, it can generate powerful negative emotions if you are made to feel excluded.

Pretus's team found that the men showed a willingness to fight and die for their sacred values, whether the Cyberball game had made them feel socially excluded or not. The men who hadn't been socially excluded weren't keen to fight and die for nonsacred values. However, the men who had been made to feel socially excluded said they were willing to fight and die for nonsacred values. If we believe what the men say, then social exclusion made them treat nonsacred values like sacred ones. As the authors noted, this was consistent with the results of surveys. When marginalized American Muslims experience discrimination, such as name-calling, racial profiling, and negative portrayals of Islam in the media, they increase their support for fundamentalist groups.[34]

The study by Pretus and colleagues also found that making men feel socially excluded caused their brains to react to nonsacred values more like sacred ones. In the whole sample of men, part of the brain called the inferior frontal gyrus increased its activity when the participants made the decision to fight for sacred values. This area was much less active when they thought about fighting for nonsacred values. Yet when the socially excluded men thought about fighting for nonsacred values, their left hemisphere inferior frontal gyrus was much more active than those of the socially included men. Thus, making the men feel socially excluded caused their brains to show the characteristic left inferior frontal gyrus activity associated with sacred values, but now in response to nonsacred values.

One function of the left inferior frontal gyrus is to help make rule-based decisions.[35] It pulls out information such as "if this, then that." It whirs into action when we look at road signs.[36] In such rule-based decision-making, costs and benefits aren't considered. In this sense, violation of sacred values (and violations of nonsacred values in the socially excluded) are signposts to spite.

———

ONCE VALUES, SACRED OR OTHERWISE, have been seen to be violated, what causes people to go on to undertake the most extreme form of costly punishment: suicide bombing? As we have seen, people typically do not want to engage in costly punishment. In the Ultimatum Game, rejecting a low offer is not what many people want to do. Many would take less costly options, such as sending a note, if they were available. Yet costly spiteful rejection is the only option Ultimatum Game players have.

Suicide bombers come to believe that all other options have failed and that violence is the only answer.[37] The terrorist organizations they are part of create this perception by framing their grievances and ideology accordingly. We saw this in the Baader-Meinhof group. They argued that talking was not possible because one couldn't reason with the generation that gave rise to Auschwitz.

Not only must a potential bomber feel that suicide bombing is the only possible answer, but they must also believe that it is a justifiable response. For this to happen, their community needs to support such an act, or at least deem it praiseworthy under specific circumstances such as martyrdom.[38]

Think back to the conversation mentioned earlier between Scott Atran and the young man whose sister had been spat at and insulted in the street. Atran told the young man that "there's always been racism" and wondered why he now planned to turn to terrorism. "Yes," said the young man, acknowledging discrimination had always existed, "but the Jihad didn't exist." Not only does the suicide bomber need to perceive unfairness, but they also have to be given a powerful supporting framework that claims spite as a legitimate response.

Consider how this happens with the Chechen Black Widows. Chechens have a norm of revenge, which is normally targeted at the person who perpetrated the wrong or at their close family. Yet due to the overwhelming force used by the Russians, Chechens' circle of revenge has widened. But why suicide bombing, which is not something most Chechens support? What has happened in Chechen society, Speckhard and Akhmedova argue, is that shattered worlds are being treated by first aid provided by a religious ideology that permits suicide bombing.[39]

The theory of shattered assumptions proposes that we all have fundamental, unarticulated assumptions about how the world works.[40] It states that we assume that the universe is just, benevolent, and predictable. It claims we take it for granted that both we and others are kind, moral, and capable, and therefore deserving of good things to happen to us. Such assumptions give life meaning and make us feel secure amidst the winds of fortune.

When a traumatic event happens, these assumptions shatter. The world becomes a cold, frightening, and unpredictable place. We realize that bad things can happen to good people. Indeed, anything could happen. We can no longer trust others. Our assumptions that we were invulnerable and in control of our lives are revealed as illusions. The resultant anxiety can be overwhelming. People need a way to cope. Some will dissociate. Others will abuse drugs. But what is really needed is a new story to make sense of the world, to cope and to live again.

In Chechen society, such a story has been provided by a religion-based terrorist ideology that resonates with a culture in which there is a duty to avenge a family member.[41] As Speckhard and Akhmedova argue, the Chechen separatist movement began as a secular one. It was then pushed by the Russian military response into accepting help from religiously oriented groups that promoted a terrorist ideology. As John Reuter puts it, Chechen suicide bombers are "desperate[,] which allows them to be deceived into being devout."[42]

So A GRIEVANCE HAS BEEN identified. The would-be bomber has been convinced that bombing is a necessary and appropriate response. But the bomber still needs to have sufficient identification with the group he is acting on behalf of to be motivated to carry out the act. Terrorists can be altruistic, and this has been argued to drive much suicide terrorism.[43] As Darwin wrote, if two groups are in conflict, the key to victory is having someone in your group who, apparently blind to alternatives, is willing to sacrifice themselves.[44]

Altruism, as we saw earlier, is the willingness to suffer costs to confer a benefit on another, such as when we donate blood or give money to charity. "Parochial altruism" is the willingness to bear a personal cost in order to harm another group, benefiting your group in the process. It is spite harnessed for explicitly altruistic ends. Altruism enriches spite, making it potentially nuclear.

Imagine playing the following laboratory game.[45] You go into the lab with a bunch of other people and are placed in one of two groups. Let's call them Team You and Team Them. The experimenter gives you ten lottery tickets and tells you there will be a prize drawing at the end of the experiment where you can win up to ten dollars. You can write your own name on up to four of the tickets. On the remaining tickets, you can either write your group's name (Team You) or the other group's name (Team Them). You then throw the tickets into a hat. When the drawing takes place, if the ticket has your name on it, you win. If it says "Team You," then members of your team share the prize. If it says "Team Them," then the other team shares the prize.

Before the drawing, you can improve your team's chances of winning. But the way you can do this makes it less likely you personally will win the prize. If you rip up a ticket with your own name on it, the experimenter will destroy five tickets marked with "Team Them." This clearly benefits your group, but it is costly to you personally. This is called "extreme parochial altruism." Your altruism is laser focused on your team. The more tickets bearing your name that you are prepared to rip up, the more extreme parochial altruism you display.

One factor that makes you more likely to perform acts of extreme parochial altruism is if you have high levels of "social-dominance orientation," a measure of the extent to which you want your own group to dominate and be superior to other groups. It is assessed by seeing how strongly people endorse statements such as "Some people are just more worthy than others," "This country would be better off if we cared less about how equal all people were," and "To get ahead in life, it is sometimes necessary to step on others."[46]

The theory behind social-dominance orientation is that societies minimize the amount of conflict within them by getting people to agree that certain groups are better than others. The superiority of a certain group then comes to be seen as a self-apparent truth. These "hierarchy-legitimizing myths" justify a society's unequal share of resources. Examples include the appalling treatment over the centuries of African Americans in the United States. Yet hierarchy-busting myths also exist; these are ideologies that explicitly do not divide persons into categories or groups. An example is *The Universal Declaration of Human Rights*, which looks to reduce social inequality.

People with higher social-dominance orientation tend to exhibit lower levels of concern for others, less support for social programs, and less engagement with protest actions. They tend to have greater levels of political and economic conservatism, nationalism, patriotism, cultural elitism, racism, sexism, and rape myth endorsement. They are more likely either to justify or to be involved in violence and illegality.[47] Politicians can target such groups. People high in social-dominance orientation were more likely to have supported Donald Trump for president.[48]

For extreme parochial altruism to drive people to act in the group's interest, people need to have a fundamental bond with their group, indeed, to be fused with their group. This is called "identity fusion." We saw earlier that it can happen between individuals, such as twins. You can also fuse your identity with a group. You become the group and the group becomes you. The resulting sense of oneness with the group creates a collective sense of invincibility and destiny.[49] Any attack on or unfairness toward the group is felt to be an attack on you. The more you feel fused with your group, the more likely you are to say that you will fight and die to defend it.[50] If your group represents a sacred value, this can produce a willingness to act spitefully to the extent of killing yourself.[51]

You may fuse with another because you share genes. We feel fused with our family. Indeed, fusion may have arisen to help families cooperate and sacrifice in the face of extreme threats, like attacks from other groups.[52] Yet, sharing *experiences* with others also helps us fuse with them. Take identical twins. The degree to which they feel fused with each other is not only predicted by their genetic similarity. It is also predicted by the number of shared experiences they have had with each other.[53] Shared experiences can create a new family.

Suffering together is a particularly powerful way for people to fuse their identities because it increases people's willingness to sacrifice for each other.[54] Just remembering shared suffering can boost identity fusion.[55] When fellow citizens feel like family, they are more prepared to die for their country. Part of this feeling may be because people come to share common core values with those they have suffered with. Because sharing core values with others is traditionally a signal of genetic relatedness, it may create the illusion of kinship, driving altruism.[56]

People who suffer together can create bonds with each other that are stronger even than their bonds to their families. In their study of Libyan revolutionary soldiers who fought against Gaddafi's regime in 2011, the anthropologist Harvey Whitehouse and his colleagues found extremely tight, family-like bonds between the soldiers. Nearly half the soldiers had bonds to each other that were *stronger* than their bonds to their own families.[57] Similarly, Scott Atran has found that the Kurdish Peshmerga often prioritize "Kurdeity" (their term for their commitment to their fellow Kurds and the defense of their homeland) over their own families. Atran gives the example of a Kurdish fighter who told him about a choice he faced during an Islamic State offensive that attacked his village. He could go into the village before Islamic State forces had gained control and get his family out. Alternatively, he could help shore up the frontline to stop the Islamic State from advancing. He could not do both. He chose the frontline. The choice, he said, haunted his every waking hour.[58]

A PUNISHING GOD REMOVES THE price tag from our counterdominant spite. Religious doctrines can also be a way for dominant spite to manifest, allowing one to raise one's status above others. And they can support spite in the form of suicide bombing. Such criminal acts are driven by a perceived threat to a group that the bomber's identity has become fused with, fired by moral outrage at the violation of a sacred norm that resonates with the bomber's own personal experiences and mobilized by justification from a respected network.[59] So what can be done about this?

Threats of retaliation reduce people's willingness to perform costly punishment. But how can this knowledge help prevent suicide bombing? How do you retaliate against the dead? You can't, but the state can let potential bombers know that it will retaliate against what a bomber leaves behind. There is some evidence this strategy works. Punitive house demolitions, performed by the Israeli Defense Forces against Palestinian suicide terrorists and terror operatives, have been found to cause an immediate and significant reduction in suicide attacks.[60]

In addition to the ethical issues raised by such actions, they also fail to address the underlying problems driving the suicide bombers: grievances. Indeed, as the authors of the house-demolition study note, "the complete ending of terror campaigns belongs to the political rather than the military realm." A key way to reduce suicide terrorism is to listen to, acknowledge, and address people's grievances. We may not agree with the grievances and cannot agree with the actions taken in their name, but they must be heard.

Another way to approach this problem is to address issues surrounding the involvement of sacred values. First, we must not let social exclusion elevate problematic nonsacred values to the level of sacred ones. And second, when sacred values are threatened, we must show, as Scott Atran has put it, how they can be "channelled into less belligerent paths."[61]

We have seen that appealing to people's financial interests can actually make them more likely to undertake costly punishment. So how could a deal involving sacred values be sweetened? One answer is for each party to offer to make sacrifices relating to one of their sacred values. A study found when Palestinians were told that Israel was prepared to give up what they believe is their sacred right to the West Bank, the Palestinians became more likely to accept a peace deal.[62] As the authors of the study observe, this attitude is also apparent among leaders of the two groups. They give the following examples to make their point. A Hamas leader and spokesman for the Palestinian government once stated, "In principle we have no problem with a Palestinian state encompassing all of our lands within the 1967 borders. . . . But let Israel apologize for our tragedy in 1948 and then we can talk about negotiating over our right of return to historic Palestine." Similarly, Isaac Ben-Israel, a former Israeli Air Force major general, has stated, "When we feel Hamas has recognized our right to exist as a Jewish state, then we can deal."

The *9/11 Commission Report* recommended that to solve the problem of terrorism, the United States and its friends should "stress educational and economic opportunity." However, the commission also observed that backward and repressive regimes "slip into societies that are without hope, where ambition and passion have no constructive outlet."[63] This suggests, as Scott Atran has argued, that we need an outlet for the ambition and passion of youth in more constructive areas than terrorism.

We need to support and inspire our young people to embrace struggles for prosocial causes, which need to be tied to sacred values. The young then need opportunities to fuse their identities with others engaged in the same endeavor. Movements such as Extinction Rebellion are already walking this path. Saving the planet has become a sacred cause, and thanks to people such as Greta Thunberg, there is a visible group to identify with. As part of this undertaking, we can harness our willingness to spite, harming our own short-term material interests and those of some corporations to promote the long-term interests of both us and our planet. We can bring altruism's shady relative back to serve the light.

Researchers are using economic games, similar to those that uncovered the existence of our spiteful side, to see how we can boost cooperation and help our planet. Such games don't pit one person against another. They engage the present to play with the future.

Such a game runs something like this. You play a multiround game with four other people. Each round represents a generation of humans. In round 1, you represent our current generation; in round 2, you represent our children's generation; in round 3, you represent our grandchildren's generation—and so on. The experimenter tells you that the Earth has a hundred billion trees. You each decide how many you want to cut down for your own profit. You can choose to take between zero and twenty billion. At the end of the game, you can change your trees into money. Your self-interest hence makes you motivated to cut down as many trees as you can.

Now if, at the end of a round, you and the other players have taken out fifty billion trees or less between you, the forests will regrow. When you come to play the next round (i.e., acting as the next generation), you will again have a hundred billion trees to split. However, if you take out more than fifty billion trees between you in a round, then the forests don't regrow. If you take out sixty billion, say, then the next round (generation) starts with only forty billion trees.

It is clearly in your long-term interest to cooperate and take out no more than ten billion trees each per round. But it is in your short-term interest to selfishly take out twenty billion trees, in case the other players act selfishly and you get left behind. So what do people do?

Researchers ran this game eighteen times.[64] They found that by the fourth generation, there were never a hundred billion trees left standing. Two-thirds of players acted co-operatively in each round by not taking out more than ten billion trees each. However, a hard-core minority of people acted selfishly. They each took out more than ten billion trees, pushing the overall group total of trees cut down to more than fifty billion, meaning the forests could not regenerate.

The researchers found a way to solve this problem. Rather than each individual deciding how many trees they would take out, democracy was introduced. Each of the five players had to vote for the number of trees to take out. The median number was the amount that each player took. So if the five people chose to cut down ten, ten, ten, fifteen, and twenty billion trees, the median number (the middle number if you write the five numbers out in order) is ten billion. This is how many trees everyone gets.

Introducing democracy radically changed the outcome of the game. In each of the twenty times the new version of the game was played, a hundred billion trees were left standing at the end. A cooperative majority were able to keep a selfish minority under control. A single individual could no longer destroy the world, whether it be for spiteful or selfish reasons. The world would no longer be burned. The black ball stayed in the bag. As the authors concluded, "Many citizens are ready to sacrifice for the greater good. We just need institutions that help them do so."

Eight
THE FUTURE OF SPITE

Spite, our "fourth behavior," is an important part of our nature. We can use our willingness to harm others at a cost to ourselves for both good and ill. Spite can be used both to exploit others and to resist exploitation. As long as there is injustice, we will need spite, and as long as there is spite, there will be injustice. Spite is both part of the problem and part of the solution. Understanding its origins and inner workings gives us the best chance to use it well. To leave it in the shadows is an invitation to join it there.

Spite is writ larger in some people's genomes than others'. Yet all our brains are listening for their cue to spite. As our environment becomes more competitive, and resources scarcer, the world shouts at us to spite. The spiteful person can excel in competitive situations because they are not afraid of getting ahead. The world knows it can speak to our brain through our stomach. Dietary changes twist the serotonin dials of our minds, making harming others more pleasurable. When another takes our share or harms our status, anger and disgust ensue. Empathy rolls back, and we see the other as less human. We inflict a cost on them, and it feels good. But we can't admit this to ourselves. We deceive ourselves into thinking we are acting to teach, deter, or reform the unfair. But the reality is that we just want to harm them. This is the *how* of spite.

The *why* of spite is simple. We spite because it pays. Actions that are immediately spiteful often lead to long-term benefits. Spite plus time equals selfishness. Counterdominant spite pulls down the bully, the dominator, and the tyrant. Here, spite can be a tool for justice. If we direct spite at those who harm others, our social capital grows. Others reward us with their cooperation and esteem. If we direct spite at those who harm us, we force them to place more value on our welfare. Over time, we have developed cheaper, safer forms of costly punishment, facilitated by language. We have also outsourced spite to god and the state. Now we can bite "with stolen teeth," to use a phrase from Nietzsche.

Dominant spite aims to put clear water between us and others. It will take an absolute loss to secure a relative advantage. We are happy to lose if it means others stay below us. We are happy to lose if others lose more. Such spite keeps us out of last place. It can help us thrive in competitive environments. Historically, reproductive benefits have flowed from this cutthroat instinct, yet it also has the potential for great harm.

Existential spite, our willingness to suffer to prove reason, nature, or inevitability wrong, seems gloriously tragic. Yet there may once have been wisdom in it. Today, it can function as an antidominance tool against the sophist. It can be used to create stretch goals that can help us achieve what we never thought possible. Such spite can boost creativity.

Spite came from the darkness. It does not aim to reform people to create fairness and cooperation. It seeks to harm the other and to bring about changes in dominance. Yet it can help usher us into the light. Spite is a sword of Damocles dangling over our interactions. It has made society fairer and more cooperative.

These benefits of spite come with an all too obvious price. Existential spite threatens our ability to use reason to solve the problems we face. Dominant spite may help us relatively speaking, but it will leave us atop a molehill, when we would be better off, in absolute terms, midway up a mountain. Counterdominance can turn into destructive resentment. If others have closed the door to social progress, our counterdominant side can summon a need for chaos that will seek to destroy all in its path. This can birth Bostrom's apocalyptic residual. Such people must not get their hands on black balls. This brings us back to where we started in Chapter 1: the Baader-Meinhof group.

We can look at this group again now, through the lens of what we have learned. Any lessons here will apply to similar groups, past, present, and future. We can tell a story of young people feeling locked out of society. Their desire for unavailable status motivates a need for chaos. They seek destruction from which they can arise out of society's ashes like phoenixes with newfound status. Here we have a "locked-out syndrome."

Looking at the Baader-Meinhof group through the lens of spite suggests something about Marxism: it is haunted by the specter of spite. This political philosophy is no stranger to the objection that it fails to take account of human nature. However, it notably fails to address the problem of spite. Although Marx acknowledged that the working class was not homogeneous, he denied that different parts of the working class could have conflicting interests.[1] What would he make of the poor being happy to give up money in order to keep the poorest below them? What happens to solidarity when the poor and the poorest have significantly different interests? A Marxist could argue that this phenomenon only occurs in societies where false consciousness pushes people to greatly value being better off than others. But if this is part of human nature, Marxism looks like a stretch goal.

Applying this insight to the Baader-Meinhof group, now we can tell a story of young people trying to capitalize on the reputational gains that come from undertaking the costly punishment of third parties. Marxism is eagerly taken up by many middle-class intellectuals and students. It allows them to make sacrifices to punish those (capitalists, organs of the state) who have harmed others (the workers). For them, Marxism is costly third-party punishment, a phenomenon that we have seen is associated with social approval. It allows activists to knock someone else down the social ladder yet call it an act of virtue.

This is an almost irresistible temptation. The novelist Aldous Huxley flagged its potency a century ago. In his 1921 book, *Crome Yellow*, Huxley wrote, "To be able to destroy with good conscience, to be able to behave badly and call your bad behaviour 'righteous indignation'—this is the height of psychological luxury, the most delicious of moral treats."

If such forces contributed to the actions of the Baader-Meinhof group and their contemporaries, such as the Weathermen in the United States, where are such people today? The forces that created them have not gone away. We may have created safer, cheaper forms of spite, but the urge to spite others has not disappeared. Perhaps such individuals live online, where cheap keystrokes have replaced costly bombings.

Half the people on Earth use online social media platforms, such as Facebook and Twitter. We have created worlds within a world. But these are not worlds we evolved to live in. The online world not only loosens the natural shackles of spite; it rewards it like never before. If a Machiavellian mind set out to make spite flow, it could not have done better than create social networks. They decrease the cost of spite and multiply its benefits. Social media creates a perfect storm for spite.

Online anonymity cuts a crucial real-world brake on spite. It eliminates the threat of retaliation. Released from this fear, people freely aim counterdominant spite at those who have more status or resources. They manically snort justice, burn others, and revel in the joy of destruction. It doesn't matter if the target earned their excess. If they got ahead on merit they will be hated all the more.

Even if you are not anonymous online, other features of the online world still encourage spite. First, it takes little effort to harm others, making spite cheap. In the online world, we are like the fabled martial artist who can destroy others with the mere tap of a finger. Second, any retaliatory costs can potentially be widely distributed. Thousands of other people may pile onto your attacks on the other person, by liking and retweeting them. As a result, any costs of retaliation may effectively be spread not just between tens of people, as in hunter-gatherer societies, but between thousands of people.

But perhaps the most important reason why having our identity known online encourages us to spite others relates to the benefits we have previously seen to be attached to acts of third-party costly punishment. In the online world, we can monitor a giant web of interactions between other people. We can weigh in on their interactions with each other and broadcast our response. Here, opportunities for costly third-party punishment open up on a scale never seen before. We can type something to harm someone who has offended or harmed someone else. This makes it third-party costly punishment (even if the cost is very small or is merely a risk of a cost). As we saw earlier, such punishment is typically esteemed by others. If we are not anonymous, then everyone knows who we are and we can publicly soak up that esteem, enhancing our reputation.

Call-out culture and cancel culture (shaming and boycotting people for transgressions) can be good things. They can hold people, including the powerful, to account for their actions and lead to positive social change. A willingness to act spitefully, to bear a cost to punish the wrong, can be crucial to creating positive change. Counterdominant spite can help us bring down those who need to be brought down. Yet it also threatens to pull down the industrious, the innovative, and the generous. The punishment can exceed the crime. As Jon Ronson, author of *So You've Been Publicly Shamed*, notes in relation to online backlashes, a disconnect has developed "between the severity of the crime and the gleeful savagery of the punishment."[2]

In our capitalist society, someone is always going to be above us. We have mixed feelings toward such people. We are set up to attend to them and learn their secrets, to cozy up to them and seek their protection. Yet counterdominant spite may also try to pull them down. It is an open question as to what causes us to cut down rather than pander to the high-status person. But if spite does frequently prune tall poppies, how do we progress, and what messages are we sending to people about the desirability of advancement?

A related problem arises from what we saw about the nature of punishment. Often it is exacted to boost one's own status. Dominance-seeking spite can masquerade as counterdominant spite. Here, online attacks become not just about creating equality for an oppressed group, but about making them dominant. The British journalist Douglas Murray makes this point. He argues that many human rights movements that seek equal rights, including those involving gender, race, and sexuality, have gone "through the crash barrier." "Not content with being equal, they have started to settle on unsustainable positions such as 'better,'" he argues.[3]

The evolutionary biologist Bret Weinstein makes a similar point. He argues that while the majority of those involved in social justice movements want to end oppression and live in a just society, a minority want to "turn the tables."[4] The aim of this minority, who he suggests may include the leaders, is a situation in which "those who were privileged are now subordinate and those who were, in their own minds, most oppressed, will be the most well-resourced and powerful."[5] Such a mindset, in the cases of long-oppressed groups, is entirely understandable. Yet we need to ensure that, in the long term, we do not replace domination and oppression with more of the same. Works such as *Genocide by the Oppressed* show the potential dangers here.[6]

So how do we deal with problematic online spite? We do not want to eliminate anonymity due to the benefits it provides. This pushes responsibility back onto the individual. Users need to reflect on why they are posting what they are posting. Yet we have seen how bad we are at figuring out when we are punishing to retaliate. We need to call on our peers to offer a structural constraint for problematic spite.

Our peers need to call out the caller-out who is acting from dominant spite, trying to boost their standing rather than aiding social justice. Mockery may be useful for this purpose.[7] Additionally, making it clear that such attacks are understood to be a way of a person trying to gain status, rather than serving any community purpose, should reduce the prestige of such events. The term "virtue signaling" has already been popularized. It reflects a recognition that third-party punishment is about the punisher seeking reputational gains. Yet there is no term to communicate simultaneously that such behavior is also about relative status-seeking by pushing the other down. Something like "virtue climbing" would capture this phenomenon better.

We also need to think carefully about whether our claims of high-minded morality are what they appear to be. Are we really spiting someone because we are morally outraged? Or are we doing it for more selfish reasons? In Chapter 4, we discussed the need to be skeptical of our claims about why we punish. This is part of a wider problem. Many researchers have concluded that we are moral hypocrites, driven by a strong motivation to *appear* fair and moral, yet not really bothered about *being* moral.[8]

We can see this when people talk about how outraged they are at violations of fairness. The psychologist C. Daniel Batson argues that anger at the abstract concept of a "fairness violation" isn't a real phenomenon. Instead, he suggests we are only really angered by being harmed.[9] For Batson, we talk of being outraged at unfairness because it signals that we are not interested in the harm done to us, thereby implying that we have a much purer, detached motive. Our concern seems noble and socially desirable rather than petty and personal. It signals that everyone should join in redressing the wrong. Speaking the language of moral outrage transforms personal concerns about harm into a global crusade that everyone should care about.

In reality, we may not care about such standards in the abstract. Moralizing can be a way for us to hold other people's "feet to the fire of censure."[10] At the same time, given half a chance, we try to escape from being moral ourselves by presenting a mere front of morality. Batson suggests that talk of moral outrage at violations of fairness may be a victim's way to increase the chance that other people will help them. We all need to take a good, hard look at our moralizing, online and offline, and be honest with both ourselves and others.

THE SPITEFUL PERSON IS OFTEN a disagreeable person. This is literally true. They score highly on the personality trait of disagreeableness.[11] This does not seem like a good character trait to possess, yet it is linked to a specific form of creativity.[12] Possessing higher levels of disagreeableness is associated with greater mathematical and scientific creativity.[13] It is unclear why this relation exists. It may relate to the spiteful person's tendency to excel in competitive environments. A propensity to engage in existential spite—attempting that which others say can't be done—could be involved. This personality trait is also associated with support for populists such as Trump.[14] Spiteful, disagreeable, populist-supporting people may be the ones most likely to help advance the sciences. Yet they are not encouraged in the academy. As the American mathematician Eric Weinstein has noted, this is problematic.[15]

I once saw an article in the press arguing that racist and sexist comments made by James Watson, who shared the 1962 Nobel Prize in Physiology or Medicine[16] for his discovery of the structure of DNA, revealed "a pernicious character entirely unrelated to his scientific greatness."[17] But was his character really "entirely unrelated"? What are the implications if it wasn't?

If society does gain from disagreeable people and the advances they can create, how can it do so without excusing their behavior, passing over the unjustifiable, or subscribing to an "ends justify the means" philosophy? Thankfully, there may be a way to dodge the problem. It turns out that if one's environment is supportive of creative thinking, then disagreeableness is less strongly linked to creativity.[18] Spite may be one way to get from A to B, but other routes may be both possible and preferable.

———

SPITE PROVIDES AN IMPORTANT CAVEAT to the sentiment, famously expressed by a strategist of President Bill Clinton, that "it's the economy, stupid." As the Ultimatum Game shows, we are prepared to take an economic hit. We may do this, in a charitable interpretation, to encourage another to behave better in the future. More likely, we do it to inflict harm on the unfair, the dominant, the elite. We may also do it to widen the gap between us and others and to stay off the bottom rung of society. The failure of elites to understand that the populace is driven by more than its narrow economic self-interest opens the doors to spite and to manipulative counterelites, with potentially disastrous results.

Although people will pay to destroy the deserved gains of others, undeserved gains are particularly precarious. Income and wealth inequality, in countries such as the United States, has been growing since the era of Thatcher and Reagan.[19] Money-burning studies suggest that how the 99 percent react to the wealth of the 1 percent will partly be influenced by whether such inequalities are seen as deserved or not. One doesn't need to be strongly left-leaning to recognize that significant propaganda efforts are made in Western societies to frame wealth inequalities as deserved. The tale of the American Dream functions to promote the story that anyone can be wealthy if they just work hard enough. Here, success is the deserved result of your labors.

Conversely, propaganda can also be used to frame inequalities as undeserved in order to sow discord. An example of this comes from a declassified 1944 CIA manual that described ways to sabotage the United States' enemies in the Second World War. One tip was to persuade managers of enemy companies who were sympathetic to the United States' aims to give inefficient workers undeserved promotions in order to lower morale and production.[20]

An old story depicting stereotypical American and Russian behavior illustrates how the United States encourages its population to deal with inequalities in a nonspiteful manner.[21] The story goes that an American farmer has a neighbor who just got a prize cow. A Russian farmer also has a neighbor with a prize cow. The American farmer's dream is to have a better cow than the neighbor. The Russian farmer's dream is that the neighbor's cow dies. Both the Russian and the American want to get ahead, but the Russian reduces overall wealth to do so, whereas the American increases it. In the United States, people are urged to deal with inequality by striving to become rich themselves rather than by spitefully destroying those with more.

But what happens if upward mobility becomes more difficult? There is some evidence this is happening. In the past half century, the American Dream has increasingly become an American illusion. Whereas 90 percent of children born in 1940 went on to earn more than their parents, only 50 percent of those entering the job market today will.[22] This isn't just because economic growth has fallen. It is because such growth has been unequally distributed. As a sense of unfairness spreads, the potential for spite grows.

Although stories function to justify inequalities, disparities are better protected if they are never visible in the first place. We all know that displaying wealth can lead to a spiteful backlash. Indeed, people will pay money to keep their wealth hidden from their peers.[23] Studies show that we massively underestimate how biased wealth distribution is in our country,[24] which suggests that efforts to conceal disparities are successful.

But what if the elite can neither justify nor hide their wealth? They may rightfully fear its spiteful destruction. An elite counterelite may rise up to use this spiteful resentment to democratically ride a wave of populism to power. Yet there is another, more dangerous possibility. As the journalist Edward Luce observes, "When inequality is high, the rich fear the mob."[25] In the wake of the 2007–2008 financial crisis, President Obama is reported to have explained to banking CEOs that "my administration is the only thing between you and the pitchforks."[26] In such situations, democracy, which permits spite voting, becomes increasingly threatening to the elite. We should hence be aware of elite attempts to undermine democracy.

In the nineteenth century, the philosopher John Stuart Mill proposed that "one person, one vote" be replaced with an alternative system wherein the better educated had more than one vote. We should keep our eyes open for the reemergence of similar ideas that undermine democracy. Ultimately, we need a more explicit public debate about what is equitable so that we can all buy into the solution.

━━━━━

WE NEED TO BE ABLE to control our spite. We need to be able to deploy spite when we choose to, not when it chooses to emerge. We need to be able to wield it for just causes rather than to promote injustice. A vital part of this is learning to control our anger. There will be times when we need to restrain anger, but also times when we need to stoke the flames of righteous anger.

Many philosophical and religious schools of thought have pointed at the dangers of anger.[27] The Stoic philosopher Seneca viewed anger as "the most hideous and frenzied of all the emotions" and argued that we should try to "cut out anger from our minds altogether." Similarly, many Buddhist traditions encourage us to give up anger, viewing our doing so as an important step on the road to wisdom.[28]

The Ultimatum Game gives us insights into what does and does not help us control our anger in response to unfairness. Suppressing your anger when you get a low offer doesn't affect rates of spiteful rejections. However, reappraising your emotions, by viewing offers with detached interest or trying to come up with possible reasons why someone might give you a certain offer, cuts rates of spiteful rejections in half.[29] To control spite, we should think about the variety of reasons for the other person's actions rather than just trying to push down our anger.

Another way we can overcome our anger-driven spite is by becoming more rational thinkers, which gives us greater control over our actions. In Chapter 2 we discussed the idea that rejecting unfair offers is our default setting. One way to avoid acting this way is by spending more time thinking rationally about what we should do.

Cognitive reflection is a measure of our ability to override our gut responses in favor of rational thought. People with high levels of cognitive reflection can use reason to override their inbuilt thinking biases. To see how well you can do this, have a go at the following questions (the answers are in the endnotes), taken from work by Dustin Calvillo and Jessica Burgeno.[30]

1. All flowers have petals. Roses have petals. If these two statements are true, can we conclude from them that roses are flowers?[31]
2. Jerry received both the fifteenth-highest and the fifteenth-lowest mark in the class. How many students are in the class?[32]
3. If it takes two nurses two minutes to measure the blood pressure of two patients, how many minutes would it take two hundred nurses to measure the blood pressure of two hundred patients?[33]

The more of these questions you got right, the higher your levels of cognitive reflection. When people with high levels of cognitive reflection play the Ultimatum Game, they are less likely to act spitefully by rejecting low offers.[34] Spite appears to be based on a gut response, whereas nonspitefully accepting offers is related to careful, conscious thinking. Improving our thinking can help us be less spiteful. Of course, it may also get us exploited. As we saw earlier in the work of Joseph Henrich, sometimes too much thought can be a bad thing.

Meditation also offers us a way to overcome anger and limit spite. When experienced Buddhist meditators received offers of one dollar from a twenty-dollar pot in the Ultimatum Game, they were half as likely to reject it compared to nonmeditators.[35] They achieved this by uncoupling their emotional reactions from their subsequent behavior, which made them better able to assess the offer on its own merits rather than relative to the other person's payoff. During unfair offers, meditators, compared to nonmeditators, showed more brain activity in areas related to attending to internal bodily states and the present moment, and less evidence of memory activation. They were probably focusing on the money available rather than on current or past experiences of unfairness.

Specific meditative practices can help us be less spiteful. The "four immeasurables" Buddhist meditations help people develop positive attitudes toward all, including those who are better off than us or with whom we are in conflict. The four immeasurables are compassion, happiness for others (appreciative joy), equanimity (being calm about others' fate), and loving, unselfish kindness. Undertaking "appreciative joy" meditation before playing the Ultimatum Game makes you act less spitefully. One study found that 25 percent of people who had undertaken such pregame meditation accepted all the unfair offers they received. Only 8 percent of people who had not undertaken this meditation acted this way.[36] Meditation didn't make people think the offers were less unfair. They were just more willing to accept the offers. It made people happier about the gains of others, even when those gains meant they were relatively worse off. It helped overcome the tall-poppy-cutting temptation of counterdominant spite.

Reading this, you may well be shouting, "Yes, but those meditators were being exploited!" Overcoming spite sounds lovely, but can we be too wary of using it, too tolerant? Arguably we can. We seem to be designed to let our emotions out. Not doing so can be bad for our mental and physical health.[37] Grievances need to be heard. Indeed, emotional venting can help reduce vendettas.[38] The aim of controlling one's anger should not be to eliminate it but rather the ability to choose when to deploy it wisely, releasing spite when reason tells us we need to. As Aristotle put it, "There is praise for someone who gets angry at the right things and with the right people, as well as in the right way, at the right time and for the right length of time."[39] Counterdominant spite can be understood not as a disease but as a response to exploitation.[40] It is a powerful tool we need to have in our pocket.

We need discernment. We need to be able to determine when spite is necessary and when forgiveness is appropriate. If we do not respond spitefully, it needs to be for the right reason. Here Nietzsche's idea of ressentiment is relevant. Nietzsche observed that we may forgive others not from a position of strength but because we are too afraid to do anything about their actions.[41] We regard our forgiveness as a virtue, but it is not always. It can be cowardice disguised as righteousness. Sometimes we need to be brave enough to spite the unjust.

Relatedly, the psychologist Michael McCullough and colleagues point out that there is a widespread perception that revenge is a disease,[42] which implies that forgiveness is a cure. McCullough, in challenging this idea, asks if revenge is like a cough. If so, it would have a function, and to suppress it could be counterproductive. The psychologist Pat Barclay adds that although there are plenty of well-publicized cases of revenge going too far, we tend to ignore cases where forgiveness goes too far.[43] Here, we might think of Tom Hanks's character in *Saving Private Ryan*, who lets a German soldier go, only to be killed by him at the end of the film. As Barclay observes, forgiveness gone too far is not seen as an error because we are told to approve of such behavior. There is, argues Barclay, an "optimal level of revenge and forgiveness for any situation." Too little revenge doesn't deter other people and actually encourages them to inflict more harm. Too little forgiveness stops us from repairing relationships. Finding the right amount of revenge—and the right amount of spite—is a challenging process of what Barclay notes is "brinkmanship."

Spite is a powerful tool not only to influence the behavior of other individuals but also to use against corporations that are acting in their narrow self-interest. As Daniel Kahneman and colleagues point out, profit-maximizing firms will be incentivized to be fair if customers are prepared to spite them for their unfair practices.[44] We need to be prepared to refrain from buying products we like, costing ourselves pleasure and corporations money, when we know the products are problematic. We need to view the corporation's immoral actions not simply as being expected or "the way business is done." Instead, we need to see them as violating a norm of decency or fairness, which we should expect them to uphold. Adopting that view will raise anger and disgust in us, motivating us to undertake difficult spiteful behavior.

Even Buddhism recognizes that spite can be useful. It has a concept of wrathful compassion. Lama John Makransky argues that we should not lash out with anger that is motivated by fear and aversion to protect our self and self-righteousness.[45] Instead, we should react like a loving parent whose child misbehaves. Wrathful compassion comes from love combined with the courage to confront the other person for their own sake, and to protect others from that person's greed, prejudice, hatred, fear, and self-protectiveness. It is such attributes, not the other person themselves, that need to be destroyed. If we can learn to use our spite as wrathful compassion, we will have a powerful new way to improve the lives of all beings.

ACKNOWLEDGMENTS

Without the researchers whose work has lit up the dark side of our nature, this book would not have been possible. Their names appear throughout the book and in the endnotes.

Neither would this book have been feasible without the generous support of my university, Trinity College Dublin.

I'm grateful for helpful comments from Shane O'Mara, Brendan Kelly, Patrick Forber, Christopher Boehm, Kelley Blewster, Kathleen McCully, my agent Bill Hamilton at A. M. Heath, and my editors, Sam Carter and Eric Henney.

Even given all the above, this book would not have been written were it not for discussions with, support from, and the love of R.

NOTES

Introduction

1. H. Schwarzbaum, *Studies in Jewish and World Folklore*, vol. 3 (Berlin: Walter de Gruyter and Co., 1968).

2. D. K. Marcus et al., "The Psychology of Spite and the Measurement of Spitefulness," *Psychological Assessment* 26, no. 2 (2014): 563–574.

3. A. R. Brereton, "Return-Benefit Spite Hypothesis: An Explanation for Sexual Interference in Stumptail Macaques (*Macaca arctoides*)," *Primates* 35, no. 2 (1994): 123–136; R. Gadagkar, "Can Animals Be Spiteful?," *Trends in Ecology and Evolution* 8, no. 7 (1993): 232–234.

4. R. Bshary and R. Bergmüller, "Distinguishing Four Fundamental Approaches to the Evolution of Helping," *Journal of Evolutionary Biology* 21, no. 2 (2008): 405–420.

5. A. Smith, *An Inquiry into the Nature and Causes of the Wealth of Nations* (Project Gutenberg, 2009, updated 2019 [1776]), bk. 5, chap. 1, pt. 2, www.gutenberg.org/files/3300/3300-h/3300-h.htm.

6. As cited in R. H. Frank, T. Gilovich, and D. T. Regan, "Does Studying Economics Inhibit Co-operation?," *Journal of Economic Perspectives* 7, no. 2 (1993): 159–171.

7. M. Hudík, "*Homo economicus* and *Homo stramineus*," *Prague Economic Papers* 24, no. 2 (2015): 154–172.

8. T. C. Scott-Phillips, T. E. Dickins, and S. A. West, "Evolutionary Theory and the Ultimate–Proximate Distinction in the Human Behavioral Sciences," *Perspectives on Psychological Science* 6, no. 1 (2011): 38–47.

9. J. J. Kuzdzal-Fick et al., "Exploiting New Terrain: An Advantage to Sociality in the Slime Mold *Dictyostelium discoideum*," *Behavioral Ecology* 18, no. 2 (2007): 433–447.

10. A. P. Melis and D. Semmann, "How Is Human Cooperation Different?," *Philosophical Transactions of the Royal Society B: Biological Sciences* 365, no. 1553 (2010): 2663–2674; T. David-Barrett and R. I. Dunbar, "Language as a Coordination Tool Evolves Slowly," *Royal Society Open Science* 3, no. 12 (2016): 160259.

11. As outlined by the United Nations: "Sustainable Development Goals," accessed July 22, 2020, www.un.org/sustainabledevelopment /sustainable-development-goals.

12. S. Pinker, *Enlightenment Now: The Case for Reason, Science, Humanism, and Progress* (New York: Penguin, 2018).

13. Y. Pasher, *Holocaust Versus Wehrmacht: How Hitler's "Final Solution" Undermined the German War Effort* (Lawrence: University Press of Kansas, 2005).

14. J. Rawls, *A Theory of Justice* (Cambridge, MA: Harvard University Press, 2009).

1. Ultimatums

1. P. W. Eager, *From Freedom Fighters to Terrorists: Women and Political Violence* (London: Routledge, 2016).

2. J. Becker, *Hitler's Children: The Story of the Baader-Meinhof Terrorist Gang* (London: Granada, 1978).

3. Becker, *Hitler's Children*. See also S. Aust, *The Baader-Meinhof Complex* (New York: Random House, 2008). I reached out to Herr Aust to try to clarify further details about the deaths of the Baader-Meinhof members in prison, but did not receive a reply.

4. Werner Güth, the researcher who initiated this work, tells me he was not inspired by the political events of that autumn.

5. D. K. Marcus et al., "The Psychology of Spite and the Measurement of Spitefulness," *Psychological Assessment* 26, no. 2 (2014): 563–574. I haven't used the author's actual questions, for copyright reasons, but you can read them in the paper. I obtained the figure of 5 to 10 percent not from the paper itself, but from the authors' raw data, which were very kindly provided to me by the study's lead author, Prof. David Marcus.

6. R. T. LaPiere, "Attitudes vs. Actions," *Social Forces* 13, no. 2 (1934): 230–237; M. W. Firmin, "Commentary: The Seminal Contribution of Richard LaPiere's Attitudes vs. Actions (1934) Research Study," *International Journal of Epidemiology* 39, no. 1 (2010): 18–20.

7. E. O. Kimbrough and J. P. Reiss, "Measuring the Distribution of Spitefulness," *PLoS One* 7, no. 8 (2012): e41812.

8. W. Poundstone, *Priceless: The Myth of Fair Value (and How to Take Advantage of It)* (London: Oneworld Publications, 2011).

9. J. Henrich et al., "In Search of *Homo economicus*: Experiments in 15 Small-Scale Societies," *American Economic Review* 91 (2001): 73–78; G. E. Bolton and R. Zwick, "Anonymity Versus Punishment in Ultimatum Bargaining," *Games and Economic Behavior* 10, no. 1 (1995): 95–121; R. H. Thaler, "Anomalies: The Ultimatum Game," *Journal of Economic Perspectives* 2, no. 4 (1988): 195–206; T. Yamagishi et al., "Rejection of Unfair Offers in the Ultimatum Game Is No Evidence of Strong Reciprocity," *Proceedings of the National Academy of Sciences* 109, no. 50 (2012): 20364–20368.

10. Poundstone, *Priceless*.

11. Poundstone, *Priceless*.

12. K. Jensen, J. Call, and M. Tomasello, "Chimpanzees Are Vengeful but Not Spiteful," *Proceedings of the National Academy of Sciences* 104, no. 32 (2007): 13046–13050; K. Jensen, J. Call, and M. Tomasello, "Chimpanzees Are Rational Maximizers in an Ultimatum Game," *Science* 318 (2007): 107–109.

13. I. Kaiser et al., "Theft in an Ultimatum Game: Chimpanzees and Bonobos Are Insensitive to Unfairness," *Biology Letters* 8, no. 6 (2012): 942–945.

14. There is some controversy about how chimps behave in the Ultimatum Game. Some researchers suggest that their failure to reject unfair offers is an artifact of the way the experiments have been run and that chimps actually do show some resistance to unfairness; see D. Proctor et al., "Chimpanzees Play the Ultimatum Game," *Proceedings of the National Academy of Sciences* 110, no. 6 (2013): 2070–2075. Primatologists suggest that psychologists may have underestimated chimps' cooperative abilities; see M. Suchak and F. B. de Waal, "Reply to Schmidt and Tomasello: Chimpanzees as Natural Team-Players," *Proceedings of the National Academy of Sciences* 113, no. 44 (2016): E6730. This remains to be seen.

15. J. R. Carter and M. D. Irons, "Are Economists Different, and If So, Why?," *Journal of Economic Perspectives* 5, no. 2 (1991): 171–177.

16. G. Marwell and R. E. Ames, "Economists Free Ride, Does Anyone Else?," *Journal of Public Economics* 15, no. 3 (1981): 295–310.

17. R. H. Frank, T. Gilovich, and D. T. Regan, "Does Studying Economics Inhibit Co-operation?," *Journal of Economic Perspectives* 7, no. 2 (1993): 159–171.

18. Most, of course, are delightful. That said . . . M. Fourcade, E. Ol-lion, and Y. Algan, "The Superiority of Economists," *Journal of Economic Perspectives* 29, no. 1 (2015): 89–114.

19. E. Hoffman, K. A. McCabe, and V. L. Smith, "On Expectations and the Monetary Stakes in Ultimatum Games," *International Journal of Game Theory* 25, no. 3 (1996): 289–301.

20. Poundstone, *Priceless*.

21. L. A. Cameron, "Raising the Stakes in the Ultimatum Game: Experimental Evidence from Indonesia," *Economic Inquiry* 37, no. 1 (1999): 47–59.

22. J. Henrich, S. J. Heine, and A. Norenzayan, "The Weirdest People in the World?," *Behavioral and Brain Sciences* 33, nos. 2–3 (2010): 61–83.

23. E. Watters, "We Aren't the World," *Pacific Standard*, updated June 14, 2017, https://psmag.com/social-justice/joe-henrich-weird-ultimatum -game-shaking-up-psychology-economics-53135.

24. Henrich et al., "In Search of *Homo economicus*."

25. Watters, "We Aren't the World."

26. Henrich et al., "In Search of *Homo economicus*."

27. C. K. Morewedge, T. Krishnamurti, and D. Ariely, "Focused on Fairness: Alcohol Intoxication Increases the Costly Rejection of Inequitable Rewards," *Journal of Experimental Social Psychology* 50 (2014): 15–20.

28. E. Halali, Y. Bereby-Meyer, and N. Meiran, "Between Self-Interest and Reciprocity: The Social Bright Side of Self-Control Failure," *Journal of Experimental Psychology: General* 143, no. 2 (2014): 745–754.

29. M. Muraven and R. F. Baumeister, "Self-Regulation and Depletion of Limited Resources: Does Self-Control Resemble a Muscle?," *Psychological Bulletin* 126, no. 2 (2000): 247–259.

30. Indeed, the majority of violent confrontations within tribes are caused by sexual conflict. See F. Guala, "Reciprocity: Weak or Strong? What Punishment Experiments Do (and Do Not) Demonstrate," *Behavioral and Brain Sciences* 35, no. 1 (2012): 45–59.

31. A. Markovitz, *Topless Prophet: The True Story of America's Most Successful Gentleman's Club Entrepreneur* (Detroit: AM Productions, 2009).

32. D. Hopkins, "American Directs Large 'Middle Finger' Statue at Home of Ex-Wife," *The Telegraph*, November 18, 2013, www.telegraph .co.uk/news/worldnews/northamerica/usa/10457437/American-directs -large-middle-finger-statue-at-home-of-ex-wife.html.

33. N. A. Christensen, "Aristotle on Anger, Justice and Punishment" (PhD diss., University College London, 2017).

34. E. S. Scott, "Pluralism, Parental Preference and Child Custody," *California Law Review* 80 (1992): 615–672.

35. M. Laing, "For the Sake of the Children: Preventing Reckless New Laws," *Canadian Journal of Family Law* 16 (1999): 229–283.

36. G. Gray, "The Fall of the House of Bartha," *New York*, July 21, 2006, http://nymag.com/news/features/18474/index1.html; C. Buckley, "How a Town House in N.Y. Went from Dream to Nightmare," *New York Times*, July 11, 2006, www.nytimes.com/2006/07/11/nyregion/11doctor.html; T. Liddy, "Honey, I Blew Up the House," *New York Post*, July 12, 2006, https://nypost.com/2006/07/12/honey-i-blew-up-the-house-dr-booms -wife-says-its-tragic-explosive-woe-for-ex/.

37. J. R. Johnston, "Parental Alignments and Rejection: An Empirical Study of Alienation in Children of Divorce," *Journal of the American Academy of Psychiatry and the Law Online* 31, no. 2 (2003): 158–170.

38. Scott, "Pluralism, Parental Preference and Child Custody."

39. A. J. L. Baker, "The Long-Term Effects of Parental Alienation on Adult Children: A Qualitative Research Study," *American Journal of Family Therapy* 33, no. 4 (2005): 289–302.

40. R. Dawkins, *The Selfish Gene*, 30th anniv. ed. (Oxford, UK: Oxford University Press, 2006), 148.

41. See *Shanabarger v. State*, 798 N.E.2d 210 (Ind. Ct. App. 2003), https://law.justia.com/cases/indiana/court-of-appeals/2003/10270302-jgb .html; T. Roche, "A Cold Dose of Vengeance," *Time*, July 4, 1999, http:// content.time.com/time/magazine/article/0,9171,27683,00.html; R. Calla- han, "Man Says He Fathered Baby to Kill It," Associated Press, June 28, 1999, https://apnews.com/292e020044f0c02ade7ea50156064f88.

42. N. Britten, "Depressed Mother Emma Hart Killed Five-Year-Old Son to Spite His Father," *The Telegraph*, July 31, 2008, www.telegraph.co.uk /news/uknews/2480795/Depressed-mother-Emma-Hart-killed-five-year -old-son-to-spite-his-father.html; T. Stelloh and E. Chuck, "Texas Mom Who Killed Daughters Called 'Family Meeting' Before Shootout," NBC News, November 18, 2013, www.nbcnews.com/news/us-news/texas-mom -who-killed-daughters-called-family-meeting-shootout-n599961.

43. See T. Joiner, *Myths About Suicide* (Cambridge, MA: Harvard Uni- versity Press, 2010).

44. S. Gardiner, "Joshua Ravindran Not Guilty of Murdering His Father Ravi," *Sydney Morning Herald*, August 15, 2013, www.smh.com.au /national/nsw/joshua-ravindran-not-guilty-of-murdering-his-father-ravi -20130815-2ryn8.html; S. Gardiner, "Father's Suicide May Have Been

'Spiteful,'" *Stuff*, August 16, 2013, www.stuff.co.nz/world/australia /9050701/Fathers-suicide-may-have-been-spiteful-judge.

45. F. J. Lanceley, *On-Scene Guide for Crisis Negotiators* (Washington, DC: CRC Press, 2005), 76.

46. To adapt something once written by the philosopher Martha Nussbaum. M. C. Nussbaum, *The Therapy of Desire: Theory and Practice in Hellenistic Ethics* (Princeton, NJ: Princeton University Press, 2013).

47. See P. J. Resnick, "Filicide in the United States," *Indian Journal of Psychiatry* 58, suppl. 2 (2016): S203.

48. M. Daly and M. Wilson, *Homicide* (New York: Transaction Publishers, 1988).

49. N. Kurczewski, "Lamborghini Supercars Exist Because of a 10-Lira Tractor Clutch," *Car and Driver*, November 16, 2018, www.carand driver.com/features/a25169632/lamborghini-supercars-exist-because-of-a -tractor/?.

50. T. Kim, "Warren Buffett Responds to Elon Musk's Criticism: 'I Don't Think He'd Want to Take Us On in Candy,'" CNBC, May 5, 2018, www.cnbc.com/2018/05/05/warren-buffett-responds-to-elon-musks -criticism-i-dont-think-hed-want-to-take-us-on-in-candy.html; R. Browne, "Moats and Candy: Here's What Elon Musk and Warren Buffett Are Clashing Over," CNBC, May 7, 2018, www.cnbc.com/2018/05/07/moats -and-candy-elon-musk-and-warren-buffet-clash.html.

51. A. Crippen, "CNBC Transcript: Warren Buffett's $200B Berkshire Blunder and the Valuable Lesson He Learned," CNBC, October 18, 2010, www.cnbc.com/id/39724884.

52. D. Kreps, "Hear Elon Musk's Surprise Rap Song 'RIP Harambe,'" *Rolling Stone*, March 31, 2019, www.rollingstone.com/music/music-news /elon-musk-rap-song-rip-harambe-815813/.

53. J. Carpenter and M. Rudisill, "Fairness, Escalation, Deference and Spite: Strategies Used in Labor-Management Bargaining Experiments with Outside Options," *Labour Economics* 10, no. 4 (2003): 427–442.

54. G. Le Bon, *The Crowd: A Study of the Popular Mind* (New York: Fischer, 1897), bk. 3, chap. 4.

55. J. A. Aimone, B. Luigi, and T. Stratmann, "Altruistic Punishment in Elections," working paper no. 4945, Center for Economic Studies and Ifo Institute (CESifo), August 2014, www.econstor.eu/bitstream /10419/102157/1/cesifo_wp4945.pdf.

56. M. Ames, "We, the Spiteful," *The Exiled*, January 22, 2011, http:// exiledonline.com/we-the-spiteful/.

57. I. Kuziemko, "The 'Last Place Aversion' Paradox," *Scientific American*, October 12, 2011, www.scientificamerican.com/article /occupy-wall-street-psychology/; I. Kuziemko et al., "'Last-Place Aversion': Evidence and Redistributive Implications," *Quarterly Journal of Economics* 129, no. 1 (2014): 105–149.

58. D. Sznycer et al., "Support for Redistribution Is Shaped by Compassion, Envy and Self-Interest, but Not a Taste for Fairness," *Proceedings of the National Academy of Sciences* 114, no. 31 (2017): 8420–8425.

59. N. Bostrom, "The Vulnerable World Hypothesis," *Global Policy* 10, no. 4 (2019): 455–476.

60. For a cautionary tale relating to nuclear fission, see Wikipedia, s.v. "David Hahn," last modified July 25, 2020, 17:56, https://en.wikipedia.org /wiki/David_Hahn. For an inspiring tale of nuclear fusion, see S. Worrall, "Why This 14-Year-Old Kid Built a Nuclear Reactor," *National Geographic*, July 26, 2015, www.nationalgeographic.com/news/2015/07/150726 -nuclear-reactor-fusion-science-kid-ngbooktalk/. Or perhaps just leave things well enough alone.

61. C. Eckel and P. Grossman, "Chivalry and Solidarity in Ultimatum Games," *Economic Inquiry* 39 (2001): 171–188; Marcus et al., "The Psychology of Spite and the Measurement of Spitefulness"; R. R. Lynam and K. J. Derefinko, "Psychopathy and Personality," in *Handbook of Psychopathy*, ed. C. J. Patrick (New York: Guilford Press, 2006), 133–155.

62. M. Moshagen, B. E. Hilbig, and I. Zettler, "The Dark Core of Personality," *Psychological Review* 125, no. 5 (2018): 656–688.

63. M. Ridley, *The Rational Optimist: How Prosperity Evolves* (London: Fourth Estate, 2010), 86.

64. H. Oosterbeek, R. Sloof, and G. Van De Kuilen, "Cultural Differences in Ultimatum Game Experiments: Evidence from a Meta-analysis," *Experimental Economics* 7, no. 2 (2004): 171–188.

2. Counterdominant Spite

1. C. Boehm, *Moral Origins: The Evolution of Virtue, Altruism, and Shame* (New York: Soft Skull Press, 2012).

2. Boehm, *Moral Origins*.

3. Again, more correctly, they will not tolerate men trying to dominate other men.

4. Lee, as cited in Boehm, *Moral Origins*, 44.

5. R. Wrangham, *The Goodness Paradox: The Strange Relationship Between Virtue and Violence in Human Evolution* (New York: Vintage, 2019), 279.

6. Boehm, *Moral Origins*, 66.

7. D. E. Erdal, "The Psychology of Sharing: An Evolutionary Approach" (doctoral thesis, University of St. Andrews, 2000), 44.

8. D. Erdal et al., "On Human Egalitarianism: An Evolutionary Product of Machiavellian Status Escalation?," *Current Anthropology* 35, no. 2 (1994): 175–183.

9. Erdal, "The Psychology of Sharing," 196.

10. C. Boehm, "Ancestral Hierarchy and Conflict," *Science* 336, no. 6083 (2012): 844–847.

11. J. Haidt, *The Righteous Mind: Why Good People Are Divided by Politics and Religion* (New York: Vintage, 2012).

12. If Death turns out to be a chimp, don't arm wrestle it for your life; challenge it to needlework or Nintendo instead. J. Caryl, "The Strength of Great Apes . . . ," *Mental Indigestion* (blog), April 3, 2009, https://mental indigestion.net/2009/04/03/760/. See, *The Seventh Seal* could have been improved.

13. Boehm, *Moral Origins*.

14. Wrangham, *The Goodness Paradox*.

15. Wrangham, *The Goodness Paradox*.

16. Again, this is not to say that we are not aggressive, only that we are less aggressive than some of our genetic relatives. For example, as Wrangham notes in *The Goodness Paradox* (21), while an appalling amount of domestic violence exists in human society, with 41 to 71 percent of women being beaten by a man at some time during their lives, all wild adult female chimpanzees experience regular serious beatings from males.

17. Wrangham, *The Goodness Paradox*, 277.

18. J. B. Peterson, *12 Rules for Life: An Antidote to Chaos* (New York: Random House, 2018).

19. J. L. Cook et al., "The Social Dominance Paradox," *Current Biology* 24, no. 23 (2014): 2812–2816.

20. C. Anderson, J. A. D. Hildreth, and L. Howland, "Is the Desire for Status a Fundamental Human Motive? A Review of the Empirical Literature," *Psychological Bulletin* 141, no. 3 (2015): 574–601.

21. P. R. Blue et al., "When Do Low Status Individuals Accept Less? The Interaction Between Self- and Other-Status During Resource Distribution," *Frontiers in Psychology* 7 (2016): 1667.

22. R. O. Deaner and A. V. Khera, "Monkeys Pay Per View: Adaptive Valuation of Social Images by Rhesus Macaques," *Current Biology* 15 (2005): 543–548.

23. M. Tomasello et al., "Two Key Steps in the Evolution of Human Cooperation," *Current Anthropology* 53 (2012): 673–692; N. J. Ratcliff et al., "The Allure of Status: High-Status Targets Are Privileged in Face Processing and Memory," *Personality and Social Psychology Bulletin* 37 (2011): 1003–1015; Anderson, Hildreth, and Howland, "Is the Desire for Status a Fundamental Human Motive?"

24. Erdal, "The Psychology of Sharing," 50.

25. Erdal et al., "On Human Egalitarianism."

26. See Boehm, *Moral Origins*; Erdal, "The Psychology of Sharing."

27. D. Morselli et al., "Social Dominance and Counter Dominance Orientation Scales (SDO/CDO): Testing Measurement Invariance," in *35th Annual Meetings of the International Society of Political Psychology* (Chicago, July 6–9, 2012).

28. Morselli et al., "Social Dominance and Counter Dominance Orientation Scales."

29. P. Brañas-Garza et al., "Fair and Unfair Punishers Coexist in the Ultimatum Game," *Scientific Reports* 4 (2014): 6025.

30. Taken from the comments section of D. K. Marcus, "How Spiteful Are You?," *Psychology Today*, May 23, 2014, www.psychologytoday.com/ie /blog/the-dark-side-personality/201405/how-spiteful-are-you#comments _bottom.

31. E. Fehr and U. Fischbacher, "The Nature of Human Altruism," *Nature* 425, no. 6960 (2003): 785–791.

32. D. Balliet, L. B. Mulder, and P. A. van Lange, "Reward, Punishment and Cooperation: A Meta-analysis," *Psychological Bulletin* 137, no. 4 (2011): 594–615.

33. E. Fehr, U. Fischbacher, and S. Gächter, "Strong Reciprocity, Human Cooperation and the Enforcement of Social Norms," *Human Nature* 13, no. 1 (2002): 1–25.

34. Guala, "Reciprocity: Weak or Strong?"

35. H. Gintis, "Strong Reciprocity and Human Sociality," *Journal of Theoretical Biology* 206, no. 2 (2000): 169–179.

36. S. Bowles and H. Gintis, "*Homo reciprocans*," *Nature* 415, no. 6868 (2002): 125–127.

37. K. M. Brethel-Haurwitz et al., "Is Costly Punishment Altruistic? Exploring Rejection of Unfair Offers in the Ultimatum Game in Real-World Altruists," *Scientific Reports* 6 (2016): 18974.

38. Brethel-Haurwitz et al., "Is Costly Punishment Altruistic?"

39. J. C. Cardenas, "Social Norms and Behavior in the Local Commons

as Seen Through the Lens of Field Experiments," *Environmental and Resource Economics* 48, no. 3 (2011): 451–485.

40. Wikipedia, s.v. "Milgram Experiment," accessed August 5, 2020, https://en.wikipedia.org/w/index.php?title=Milgram_experiment&oldid =971251066.

41. Brethel-Haurwitz et al., "Is Costly Punishment Altruistic?"

42. E. Fehr and K. M. Schmidt, "A Theory of Fairness, Competition and Cooperation," *Quarterly Journal of Economics* 114, no. 3 (1999): 817–868.

43. Fehr and Schmidt, "A Theory of Fairness, Competition and Cooperation."

44. S. Sul et al., "Medial Prefrontal Cortical Thinning Mediates Shifts in Other-Regarding Preferences During Adolescence," *Scientific Reports* 7, no. 1 (2017): 8510.

45. E. L. Khalil and N. Feltovich, "Moral Licensing, Instrumental Apology and Insincerity Aversion: Taking Immanuel Kant to the Lab," *PLoS One* 13, no. 11 (2018): e0206878.

46. A. Marchetti et al., "Expectations and Outcome: The Role of Proposer Features in the Ultimatum Game," *Journal of Economic Psychology* 32, no. 3 (2011): 446–449.

47. S. Harris, "Can We Build AI Without Losing Control Over It?," TEDSummit, June 2016, https://www.ted.com/talks/sam_harris_can_we _build_ai_without_losing_control_over_it?language=en.

48. S. Blount, "When Social Outcomes Aren't Fair: The Effect of Causal Attributions on Preferences," *Organizational Behavior and Human Decision Processes* 63, no. 2 (1995): 131–144. See also A. G. Sanfey et al., "The Neural Basis of Economic Decision-Making in the Ultimatum Game," *Science* 300, no. 5626 (2003): 1755–1758.

49. Henrich et al., "In Search of *Homo economicus.*"

50. P. Vavra, L. J. Chang, and A. G. Sanfey, "Expectations in the Ultimatum Game: Distinct Effects of Mean and Variance of Expected Offers," *Frontiers in Psychology* 9 (2018): 992.

51. A. G. Sanfey, "Expectations and Social Decision-Making: Biasing Effects of Prior Knowledge on Ultimatum Responses," *Mind and Society* 8 (2009): 93–107.

52. D. J. Zizzo and A. J. Oswald, "Are People Willing to Pay to Reduce Others' Incomes?," *Annales d'Economie et de Statistique* (2001): 39–65; D. J. Zizzo, "Money Burning and Rank Egalitarianism with Random Dictators," *Economics Letters* 81, no. 2 (2003): 263–266.

53. K. Jensen et al., "What's in It for Me? Self-Regard Precludes Altruism and Spite in Chimpanzees," *Proceedings of the Royal Society B: Biological Sciences* 273, no. 1589 (2006): 1013–1021.

54. D. J. de Quervain et al., "The Neural Basis of Altruistic Punishment," *Science* 305, no. 5688 (2004): 1254–1258.

55. Henrich et al., "In Search of *Homo economicus*."

56. J. Rodrigues et al., "Altruistic Punishment Is Connected to Trait Anger, Not Trait Altruism, if Compensation Is Available," *Heliyon* 4, no. 11 (2018): e00962; E. C. Seip, W. W. Van Dijk, and M. Rotteveel, "Anger Motivates Costly Punishment of Unfair Behavior," *Motivation and Emotion* 38, no. 4 (2014): 578–588.

57. K. Gospic et al., "Limbic Justice: Amygdala Involvement in Immediate Rejection in the Ultimatum Game," *PLoS Biology* 9, no. 5 (2011): e1001054.

58. K. S. Birditt and K. L. Fingerman, "Age and Gender Differences in Adults' Descriptions of Emotional Reactions to Interpersonal Problems," *Journals of Gerontology Series B: Psychological Sciences and Social Sciences* 58, no. 4 (2003): P237–P245.

59. H. A. Chapman et al., "In Bad Taste: Evidence for the Oral Origins of Moral Disgust," *Science* 323, no. 5918 (2009): 1222–1226.

60. A. G. Sanfey et al., "The Neural Basis of Economic Decision-Making in the Ultimatum Game," *Science* 300, no. 5626 (2003): 1755–1758.

61. J. M. Salerno and L. C. Peter-Hagene, "The Interactive Effect of Anger and Disgust on Moral Outrage and Judgments," *Psychological Science* 24, no. 10 (2013): 2069–2078.

62. T. Yamagishi et al., "The Private Rejection of Unfair Offers and Emotional Commitment," *Proceedings of the National Academy of Sciences* 106, no. 28 (2009): 11520–11523.

63. Gospic et al., "Limbic Justice." To take another example, oxytocin reduces aggression in males, but not females. As you would predict, giving oxytocin to people playing the Ultimatum Game reduces men's likelihood of spitefully rejecting low offers, but not women's. See R. Zhu et al., "Intranasal Oxytocin Reduces Reactive Aggression in Men but Not in Women: A Computational Approach," *Psychoneuroendocrinology* 108 (2019): 172–181.

64. V. Grimm and F. Mengel, "Let Me Sleep on It: Delay Reduces Rejection Rates in Ultimatum Games," *Economics Letters* 111, no. 2 (2011): 113–115.

65. B. D. Dunn et al., "Gut Feelings and the Reaction to Perceived Inequity: The Interplay Between Bodily Responses, Regulation and Perception

Shapes the Rejection of Unfair Offers on the Ultimatum Game," *Cognitive and Affective Behavioral Neuroscience* 12 (2012): 419–429.

66. G. Gilam et al., "Attenuating Anger and Aggression with Neuromodulation of the vmPFC: A Simultaneous tDCS-fMRI Study," *Cortex* 109 (2018): 156–170; G. Gilam et al., "The Anger-Infused Ultimatum Game: A Reliable and Valid Paradigm to Induce and Assess Anger," *Emotion* 19, no. 1 (2019): 84–96.

67. Given that neurostimulation led people to perceive low offers as less unfair, the stimulation could have had its effects not by helping people control their anger, but by changing their fairness perceptions. Then again, perhaps because they felt less anger, they inferred that the offer was less unfair. This remains to be seen.

68. T. Baumgartner et al., "Dorsolateral and Ventromedial Prefrontal Cortex Orchestrate Normative Choice," *Nature Neuroscience* 14 (2011): 1468–1474.

69. D. Knoch et al., "Diminishing Reciprocal Fairness by Disrupting the Right Prefrontal Cortex," *Science* 314, no. 5800 (2006): 829–832.

70. D. Knoch et al., "Disrupting the Prefrontal Cortex Diminishes the Human Ability to Build a Good Reputation," *Proceedings of the National Academy of Sciences* 106, no. 49 (2009): 20895–20899.

71. C. Speitel, E. Traut-Mattausch, and E. Jonas, "Functions of the Right DLPFC and Right TPJ in Proposers and Responders in the Ultimatum Game," *Social Cognitive and Affective Neuroscience* 14, no. 3 (2019): 263–270.

72. D. Fetchenhauer and X. Huang, "Justice Sensitivity and Distributive Decisions in Experimental Games," *Personality and Individual Differences* 36, no. 5 (2004): 1015–1029.

73. T. Singer et al., "Empathic Neural Responses Are Modulated by the Perceived Fairness of Others," *Nature* 439, no. 7075 (2006): 466–469.

74. Notably, this effect was only found in men, suggesting they may be set up to punish more easily than women, at least in response to physical threats.

75. T. S. Rai, P. Valdesolo, and J. Graham, "Dehumanization Increases Instrumental Violence, but Not Moral Violence," *Proceedings of the National Academy of Sciences* 114, no. 32 (2017): 8511–8516.

76. K. M. Fincher and P. E. Tetlock, "Perceptual Dehumanization of Faces Is Activated by Norm Violations and Facilitates Norm Enforcement," *Journal of Experimental Psychology: General* 145, no. 2 (2016): 131–146.

77. D. Ewing, V. Zeigler-Hill, and J. Vonk, "Spitefulness and Deficits in the Social-Perceptual and Social-Cognitive Components of Theory of Mind," *Personality and Individual Differences* 91 (2016): 7–13.

78. B. Bryson, *Notes from a Small Island* (London: Transworld, 1995), 336.

79. This point is made by E. Seip, W. Vandijk, and M. Rotteveel, "On Hotheads and Dirty Harries," *Annals of the New York Academy of Sciences* 1167, no. 1 (2009): 190–196.

80. E. Fehr and U. Fischbacher, "Third-Party Punishment and Social Norms," *Evolution and Human Behavior* 25, no. 2 (2004): 63–87.

81. K. Riedl et al., "No Third-Party Punishment in Chimpanzees," *Proceedings of the National Academy of Sciences* 109, no. 37 (2012): 14824–14829.

82. K. McAuliffe, J. J. Jordan, and F. Warneken, "Costly Third-Party Punishment in Young Children," *Cognition* 134 (2015): 1–10.

83. E. J. Pedersen, R. Kurzban, and M. E. McCullough, "Do Humans Really Punish Altruistically? A Closer Look," *Proceedings of the Royal Society B: Biological Sciences* 280, no. 1758 (2013): 20122723.

84. Pedersen, Kurzban, and McCullough, "Do Humans Really Punish Altruistically?"

85. M. D. Santos, D. J. Rankin, and C. Wedekind, "The Evolution of Punishment Through Reputation," *Proceedings of the Royal Society B: Biological Sciences* 278, no. 1704 (2011): 371–377.

86. A. Sell, "Recalibration Theory of Anger," *Encyclopedia of Evolutionary Psychological Science* (2017): 1–3.

87. J. Tooby and L. Cosmides, "The Evolutionary Psychology of the Emotions and Their Relationship to Internal Regulatory Variables," in *Handbook of Emotions*, ed. M. Lewis, J. M. Haviland-Jones, and L. F. Barrett (London: Guilford Press, 2008), 114–137.

88. M. E. McCullough, R. Kurzban, and B. A. Tabak, "Cognitive Systems for Revenge and Forgiveness," *Behavioral and Brain Sciences* 36, no. 1 (2013): 1–15.

89. N. J. Raihani and R. Bshary, "Third-Party Punishers Are Rewarded, but Third-Party Helpers Even More So," *Evolution* 69, no. 4 (2015): 993–1003.

90. K. Sylwester and G. Roberts, "Reputation-Based Partner Choice Is an Effective Alternative to Indirect Reciprocity in Solving Social Dilemmas," *Evolution and Human Behavior* 34, no. 3 (2013): 201–226.

91. A. Dreber et al., "Winners Don't Punish," *Nature* 452, no. 7185 (2008): 348–351.

92. J. Heffner and O. FeldmanHall, "Why We Don't Always Punish: Preferences for Non-punitive Responses to Moral Violations," *Scientific Reports* 9, no. 1 (2019): 1–13.

93. P. Barclay, "Reputational Benefits for Altruistic Punishment," *Evolution and Human Behavior* 27, no. 5 (2006): 325–344.

94. Heffner and FeldmanHall, "Why We Don't Always Punish."

95. As noted by Guala in "Reciprocity: Weak or Strong?"

96. L. Molleman et al., "People Prefer Coordinated Punishment in Cooperative Interactions," *Nature Human Behavior* 3, no. 11 (2019): 1145–1153.

97. J. Berger and D. Hevenstone, "Norm Enforcement in the City Revisited: An International Field Experiment of Altruistic Punishment, Norm Maintenance, and Broken Windows," *Rationality and Society* 28, no. 3 (2016): 299–319.

98. E. A. Hoebel, *The Law of Primitive Man: A Study in Comparative Legal Dynamics* (Cambridge, MA: Harvard University Press, 1954).

99. Guala, "Reciprocity: Weak or Strong?"

100. N. Henrich and J. P. Henrich, *Why Humans Cooperate: A Cultural and Evolutionary Explanation* (Oxford, UK: Oxford University Press, 2007).

101. Three, if you consider that without it we would not have had the TV series *Gossip Girl*.

102. M. Feinberg et al., "The Virtues of Gossip: Reputational Information Sharing as Pro-social Behavior," *Journal of Personality and Social Psychology* 102, no. 5 (2012): 1015–1030; J. Wu, D. Balliet, and P. A. van Lange, "When Does Gossip Promote Generosity? Indirect Reciprocity Under the Shadow of the Future," *Social Psychological and Personality Science* 6, no. 8 (2015): 923–930; E. Jolly and L. J. Chang, "Gossip Drives Vicarious Learning and Facilitates Robust Social Connections," *PsyArXiv* (2018), https://doi.org/10.31234/osf.io/qau5s.

103. J. Wu, D. Balliet, and P. A. van Lange, "Gossip Versus Punishment: The Efficiency of Reputation to Promote and Maintain Cooperation," *Scientific Reports* 6 (2016): 23919.

104. M. M. Turner et al., "Relational Ruin or Social Glue? The Joint Effect of Relationship Type and Gossip Valence on Liking, Trust and Expertise," *Communication Monographs* 70 (2003): 129–141.

105. E. Xiao and D. Houser, "Emotion Expression in Human Punishment Behavior," *Proceedings of the National Academy of Sciences* 102, no. 20 (2005): 7398–7401.

106. D. Masclet et al., "Monetary and Nonmonetary Punishment in the Voluntary Contributions Mechanism," *American Economic Review* 93, no. 1 (2003): 366–380.

107. L. Balafoutas, N. Nikiforakis, and B. Rockenbach, "Altruistic Punishment Does Not Increase with the Severity of Norm Violations in the Field," *Nature Communications* 7, no. 1 (2016): 1–6.

108. N. J. Raihani and R. Bshary, "Punishment: One Tool, Many Uses," *Evolutionary Human Sciences* 1 (2019): e12.

109. D. Baldassarri and G. Grossman, "Centralized Sanctioning and Legitimate Authority Promote Cooperation in Humans," *Proceedings of the National Academy of Sciences* 108 (2011): 11023–11027.

110. N. J. Raihani and K. McAuliffe, "Human Punishment Is Motivated by Inequity Aversion, Not a Desire for Reciprocity," *Biology Letters* 8, no. 5 (2012): 802–804.

111. B. Herrmann, C. Thöni, and S. Gächter, "Antisocial Punishment Across Societies," *Science* 319, no. 5868 (2008): 1362–1367.

112. C. Boehm, *Hierarchy in the Forest: The Evolution of Egalitarian Behavior* (Cambridge, MA: Harvard University Press, 1999), as cited in A. Pleasant and P. Barclay, "Why Hate the Good Guy? Antisocial Punishment of High Cooperators Is Greater When People Compete to Be Chosen," *Psychological Science* 29, no. 6 (2018): 868–876.

113. M. Foucault, *The Courage of Truth: Lectures at the College de France 1983–1984*, trans. Graham Burchell (New York: Palgrave Macmillan, 2011).

114. J. A. Minson and B. Monin, "Do-Gooder Derogation: Disparaging Morally Motivated Minorities to Defuse Anticipated Reproach," *Social Psychological and Personality Science* 3, no. 2 (2012): 200–207.

115. Pleasant and Barclay, "Why Hate the Good Guy?"

116. Pleasant and Barclay, "Why Hate the Good Guy?"

117. Pleasant and Barclay, "Why Hate the Good Guy?"

118. P. Barclay, "Strategies for Cooperation in Biological Markets, Especially for Humans," *Evolution and Human Behavior* 34 (2013): 164–175.

119. Herrmann, Thöni, and Gächter, "Antisocial Punishment Across Societies."

120. Brañas-Garza et al., "Fair and Unfair Punishers Coexist in the Ultimatum Game."

121. R. Niebuhr, *Moral Man and Immoral Society: A Study in Ethics and Politics* (Louisville, KY: Westminster John Knox Press, 2013 [1932]).

3. Dominant Spite

1. I. Kuziemko, "The 'Last Place Aversion' Paradox," *Scientific American*, October 12, 2011, www.scientificamerican.com/article/occupy-wall-street-psychology/; I. Kuziemko et al., "'Last-Place Aversion': Evidence and Redistributive Implications," *Quarterly Journal of Economics* 129, no. 1 (2014): 105–149.

2. B. Herrmann and H. Orzen, "The Appearance of *Homo rivalis*: Social Preferences and the Nature of Rent Seeking," (CeDEx discussion paper series, No. 2008–10, 2008). These researchers do not suggest we create a new type of person called *Homo rivalis* but are keen to stress that most of us can act like this, under the right (or wrong) conditions. Our nature is somewhat fluid and is shaped by the conditions of society.

3. Hat tip to John Milton's *Paradise Lost*.

4. P. A. van Lange et al., "Development of Prosocial, Individualistic, and Competitive Orientations: Theory and Preliminary Evidence," *Journal of Personality and Social Psychology* 73, no. 4 (1997): 733–746.

5. I'm pulling these data from Study 4 of van Lange et al., "Development of Prosocial, Individualistic, and Competitive Orientations," which surveyed the general population and assessed social value orientation in 1,728 people. One hundred thirty-five couldn't be classified, leaving 1,593 people. Of these, 1,134 (66 percent) were classified as prosocial, 340 (20 percent) as individualistic, and 119 (7 percent) as competitive. I worked out the percentages based not on the 1,593 but on the overall sample of 1,728 people, allowing that some (8 percent) were unclassifiable.

6. A. Falk, E. Fehr, and U. Fischbacher, "Driving Forces Behind Informal Sanctions," *Econometrica* 73, no. 6 (2005): 2017–2030.

7. D. Houser and E. Xiao, "Inequality-Seeking Punishment," *Economics Letters* 109, no. 1 (2010): 20–23.

8. K. Abbink and A. Sadrieh, "The Pleasure of Being Nasty," *Economics Letters* 105, no. 3 (2009): 306–308.

9. N. Steinbeis and T. Singer, "The Effects of Social Comparison on Social Emotions and Behavior During Childhood: The Ontogeny of Envy and Schadenfreude Predicts Developmental Changes in Equity-Related Decisions," *Journal of Experimental Child Psychology* 115, no. 1 (2013): 198–209.

10. It could be objected that people weren't technically acting spitefully in this game, because they didn't pay a price to harm the other. However, when the game is played and you have to pay your hard-earned money to destroy the other player's money (as opposed to being able to do it for free),

25 percent of people still pay money to destroy the other person's money anonymously.

11. A. Rustichini and A. Vostroknutov, "Competition with Skill and Luck," ResearchGate, 2008, www.researchgate.net/publication /228372140_Competition_with_Skill_and_Luck.

12. P. Barclay and B. Stoller, "Local Competition Sparks Concerns for Fairness in the Ultimatum Game," *Biology Letters* 10, no. 5 (2014): 20140213.

13. S. E. Hill and D. M. Buss, "Envy and Positional Bias in the Evolutionary Psychology of Management," *Managerial and Decision Economics* 27, no. 2–3 (2006): 131–143.

14. A. Gardner and S. A. West, "Spite and the Scale of Competition," *Journal of Evolutionary Biology* 17, no. 6 (2004): 1195–1203.

15. S. Prediger, B. Vollan, and B. Herrmann, "Resource Scarcity and Antisocial Behavior," *Journal of Public Economics* 119 (2014): 1–9.

16. N. J. Raihani and R. Bshary, "Punishment: One Tool, Many Uses," *Evolutionary Human Sciences* 1 (2019): e12.

17. M. J. Crockett et al., "Serotonin Modulates Behavioral Reactions to Unfairness," *Science* 320, no. 5884 (2008): 1739.

18. M. J. Crockett et al. "Serotonin Selectively Influences Moral Judgment and Behavior Through Effects on Harm Aversion," *Proceedings of the National Academy of Sciences* 107, no. 40 (2010): 17433–17438.

19. M. J. Crockett et al., "Dissociable Effects of Serotonin and Dopamine on the Valuation of Harm in Moral Decision Making," *Current Biology* 25, no. 14 (2015): 1852–1859; M. J. Crockett et al., "Serotonin Modulates Striatal Responses to Fairness and Retaliation in Humans," *Journal of Neuroscience* 33 (2013): 3505–3513.

20. Crockett and colleagues also noted that reductions in serotonin are associated, in both humans and primates, with increases in uncontrolled aggression. Such aggression often leads to primates getting badly hurt or even killed. It may also facilitate acts of spite.

21. M. E. McCullough et al., "Harsh Childhood Environmental Characteristics Predict Exploitation and Retaliation in Humans," *Proceedings of the Royal Society B: Biological Sciences* 280, no. 1750 (2013): 20122104.

22. T. C. Burnham, "High-Testosterone Men Reject Low Ultimatum Game Offers," *Proceedings of the Royal Society B: Biological Sciences* 274, no. 1623 (2007): 2327–2330.

23. M. L. Batrinos, "Testosterone and Aggressive Behavior in Man," *International Journal of Endocrinology and Metabolism* 10, no. 3 (2012): 563–568.

24. G. Nave et al., "Single-Dose Testosterone Administration Increases Men's Preference for Status Goods," *Nature Communications* 9, no. 1 (2018): 1–8.

25. L. Balafoutas, R. Kerschbamer, and M. Sutter, "Distributional Preferences and Competitive Behavior," *Journal of Economic Behavior and Organization* 83, no. 1 (2012): 125–135.

4. Spite, Evolution, and Punishment

1. B. Wallace et al., "Heritability of Ultimatum Game Responder Behavior," *Proceedings of the National Academy of Sciences* 104, no. 40 (2007): 15631–15634.

2. S. Zhong et al., "Dopamine D4 Receptor Gene Associated with Fairness Preference in Ultimatum Game," *PLoS One* 5, no. 11 (2010); M. Reuter et al., "The Influence of Dopaminergic Gene Variants on Decision Making in the Ultimatum Game," *Frontiers in Human Neuroscience* 7 (2013): 242.

3. R. Dawkins, *A Devil's Chaplain: Selected Writings* (London: Weidenfeld and Nicolson, 2003), 171.

4. A. Vázquez et al., "Sharing Genes Fosters Identity Fusion and Altruism," *Self and Identity* 16, no. 6 (2017): 684–702.

5. Broadly speaking, though, the Wilsonian focus on the benefits of spite to your kin and the Hamiltonian focus on harm to your nonkin are ultimately two sides of the same coin; L. Lehmann, K. Bargum, and M. Reuter, "An Evolutionary Analysis of the Relationship Between Spite and Altruism," *Journal of Evolutionary Biology* 19, no. 5 (2006): 1507–1516.

6. R. Smead and P. Forber, "The Evolutionary Dynamics of Spite in Finite Populations," *Evolution: International Journal of Organic Evolution* 67, no. 3 (2013): 698–707; A. Gardner and S. A. West, "Spite and the Scale of Competition," *Journal of Evolutionary Biology* 17, no. 6 (2004): 1195–1203.

7. L. Keller and K. G. Ross, "Selfish Genes: A Green Beard in the Red Fire Ant," *Nature* 394, no. 6693 (1998): 573.

8. The question that probably occurs to you is why version B of the gene isn't completely destroyed, leaving only version A in the population. The answer is that if you have a particularly strong version A, you die early. The low probability of such counterpressure existing probably explains why greenbeard genes are quite rare, with most being pushed into a single version over time that is possessed by all members of a species.

9. S. A. West and A. Gardner, "Altruism, Spite and Greenbeards," *Science* 327, no. 5971 (2010): 1341–1344; A. Gardner et al., "Spiteful Soldiers

and Sex Ratio Conflict in Polyembryonic Parasitoid Wasps," *American Naturalist* 169, no. 4 (2007): 519–533.

10. A. Gardner and S. A. West, "Spite," *Current Biology* 16, no. 17 (2006): R662–R664.

11. A. Bhattacharya et al., "Evolution of Increased Virulence Is Associated with Decreased Spite in the Insect-Pathogenic Bacterium *Xenorhabdus nematophila*," *Biology Letters* 15, no. 8 (2019): 20190432.

12. M. Hauser, K. McAuliffe, and P. R. Blake, "Evolving the Ingredients for Reciprocity and Spite," *Philosophical Transactions of the Royal Society B: Biological Sciences* 364, no. 1533 (2009): 3255–3266.

13. K. Jensen, "Punishment and Spite, the Dark Side of Cooperation," *Philosophical Transactions of the Royal Society B: Biological Sciences* 365, no. 1553 (2010): 2635–2650.

14. R. Gadagkar, "Can Animals Be Spiteful?," *Trends in Ecology and Evolution* 8, no. 7 (1993): 232–234.

15. Gadagkar, "Can Animals Be Spiteful?" Also see Jensen, "Punishment and Spite, the Dark Side of Cooperation."

16. A. R. Brereton, "Return-Benefit Spite Hypothesis: An Explanation for Sexual Interference in Stumptail Macaques (*Macaca arctoides*)," *Primates* 35, no. 2 (1994): 123–136; R. Trivers, *Social Evolution* (Menlo Park, CA: Benjamin-Cummings, 1985); Jensen, "Punishment and Spite, the Dark Side of Cooperation."

17. Brereton, "Return-Benefit Spite Hypothesis."

18. Hauser, McAuliffe, and Blake, "Evolving the Ingredients for Reciprocity and Spite."

19. R. A. Johnstone and R. Bshary, "Evolution of Spite Through Indirect Reciprocity," *Proceedings of the Royal Society B: Biological Sciences* 271, no. 1551 (2004): 1917–1922.

20. P. Forber and R. Smead, "The Evolution of Fairness Through Spite," *Proceedings of the Royal Society B: Biological Sciences* 281, no. 1780 (2014): 20132439.

21. F. Nietzsche, *On the Genealogy of Morality*, ed. K. Ansell-Pearson, trans. C. Diethe (Cambridge, UK: Cambridge University Press, 2006 [1887]), 20.

22. Which brings to mind a classic scene from *Blackadder Goes Forth*: www.youtube.com/watch?v=yZT-wVnFn60.

23. X. Chen, A. Szolnoki, and M. Perc, "Probabilistic Sharing Solves the Problem of Costly Punishment," *New Journal of Physics* 16, no. 8 (2014): 083016.

24. N. J. Raihani and R. Bshary, "Punishment: One Tool, Many Uses," *Evolutionary Human Sciences* 1 (2019): e12.

25. Raihani and McAuliffe, "Human Punishment Is Motivated by Inequity Aversion."

26. D. Balliet, L. B. Mulder, and P. A. van Lange, "Reward, Punishment and Cooperation: A Meta-analysis," *Psychological Bulletin* 137, no. 4 (2011): 594–615.

27. Raihani and Bshary, "Punishment."

28. A. Dreber et al., "Winners Don't Punish," *Nature* 452, no. 7185 (2008): 348–351.

29. D. G. Rand and M. A. Nowak, "The Evolution of Antisocial Punishment in Optional Public Goods Games," *Nature Communications* 2, no. 1 (2011): 1–7.

30. Dreber et al., "Winners Don't Punish."

31. M. J. Crockett, Y. Özdemir, and E. Fehr, "The Value of Vengeance and the Demand for Deterrence," *Journal of Experimental Psychology: General* 143, no. 6 (2014): 2279–2286.

32. Crockett, Özdemir, and Fehr, "The Value of Vengeance and the Demand for Deterrence."

33. E. L. Khalil and N. Feltovich, "Moral Licensing, Instrumental Apology and Insincerity Aversion: Taking Immanuel Kant to the Lab," *PLoS One* 13, no. 11 (2018): e0206878.

34. E. Xiao and D. Houser, "Emotion Expression in Human Punishment Behavior," *Proceedings of the National Academy of Sciences* 102, no. 20 (2005): 7398–7401.

5. Spite and Freedom

1. Rousseau, as cited in I. Berlin, *Four Essays on Liberty* (Oxford, UK: Oxford University Press, 1969).

2. S. Pinker, *Enlightenment Now: The Case for Reason, Science, Humanism, and Progress* (New York: Penguin, 2018), 8.

3. Wikipedia, s.v. "Milgram Experiment," accessed August 5, 2020, https://en.wikipedia.org/w/index.php?title=Milgram_experiment&oldid=971251066.

4. To get a sense of this, you can see the illusionist Derren Brown's re-creation of the experiment here: www.youtube.com/watch?v=Xxq4Q tK3j0Y.

5. J. M. Burger, Z. M. Girgis, and C. C. Manning, "In Their Own Words: Explaining Obedience to Authority Through an Examination of

Participants' Comments," *Social Psychological and Personality Science* 2, no. 5 (2011): 460–466.

6. R. M. Ryan and E. L. Deci, "Self-Determination Theory and the Facilitation of Intrinsic Motivation, Social Development and Well-Being," *American Psychologist* 55, no. 1 (2000): 68–78.

7. M. Kühler and N. Jelinek, eds., *Autonomy and the Self*, vol. 118 (Berlin: Springer Science and Business Media, 2012).

8. A. MacIntyre, *After Virtue: A Study in Moral Theory* (London: Bloomsbury, 2011); J. B. Schneewind, *The Invention of Autonomy: A History of Modern Moral Philosophy* (Cambridge, UK: Cambridge University Press, 1998); L. Siedentop, *Inventing the Individual: The Origins of Western Liberalism* (Cambridge, MA: Harvard University Press, 2014).

9. R. M. Ryan and E. L. Deci, "Self-Regulation and the Problem of Human Autonomy: Does Psychology Need Choice, Self-Determination and Will?," *Journal of Personality* 74, no. 6 (2006): 1557–1586.

10. R. Johnson and A. Cureton, "Kant's Moral Philosophy," *Stanford Encyclopedia of Philosophy*, ed. E. N. Zalta, Spring 2019 edition, https://plato.stanford.edu/archives/spr2019/entries/kant-moral/.

11. J. Christman and J. Anderson, eds., *Autonomy and the Challenges to Liberalism: New Essays* (Cambridge, UK: Cambridge University Press, 2005).

12. Hat tip to Winston Churchill and his view on democracy.

13. J. W. Brehm, *A Theory of Psychological Reactance* (New York: Academic Press, 1966).

14. C. H. Miller et al., "Identifying Principal Risk Factors for the Initiation of Adolescent Smoking Behaviors: The Significance of Psychological Reactance," *Health Communication* 19, no. 3 (2006): 241–252.

15. K. D. Vohs and J. W. Schooler, "The Value of Believing in Free Will: Encouraging a Belief in Determinism Increases Cheating," *Psychological Science* 19, no. 1 (2008): 49–54.

16. R. F. Baumeister, E. J. Masicampo, and C. N. DeWall, "Pro-social Benefits of Feeling Free: Disbelief in Free Will Increases Aggression and Reduces Helpfulness," *Personality and Social Psychology Bulletin* 35, no. 2 (2009): 260–268.

17. J. M. Twenge, L. Zhang, and C. Im, "It's Beyond My Control: A Cross-Temporal Meta-analysis of Increasing Externality in Locus of Control, 1960–2002," *Personality and Social Psychology Review* 8, no. 3 (2004): 308–319.

18. S. McCarthy-Jones, "The 'Braveheart Effect'—and How Companies Manipulate Our Desire for Freedom," *The Conversation*,

August 31, 2018, https://theconversation.com/the-braveheart-effect-and
-how-companies-manipulate-our-desire-for-freedom-102057.

19. Brehm, *A Theory of Psychological Reactance*.

20. G. E. Lenehan and P. O'Neill, "Reactance and Conflict as Deter-
minants of Judgment in a Mock Jury Experiment," *Journal of Applied Social
Psychology* 11, no. 3 (1981): 231–239.

21. T. E. Hannah, E. R. Hannah, and B. Wattie, "Arousal of Psychological Reactance as a Consequence of Predicting an Individual's Behavior,"
Psychological Reports 37, no. 2 (1975): 411–420.

22. S. Worchel and J. W. Brehm, "Effect of Threats to Attitudinal
Freedom as a Function of Agreement with the Communicator," *Journal of
Personality and Social Psychology* 14, no. 1 (1970): 18–22.

23. M. Mihajlov, "Life = Freedom: The Symbolism of 2×2 = 4 in
Dostoevsky, Zamyatin and Orwell," *Crisis Magazine*, orig. pub. October
1, 1984, accessed August 22, 2020, www.crisismagazine.com/1984/life
-freedom-the-symbolism-of-2x2-4-in-dostoevsky-zamyatin-orwell.

24. Unless otherwise indicated, all Dostoyevsky quotations in this
chapter come from one of the following three sources: F. Dostoyevsky,
Notes from the Underground (Project Gutenberg, 2008), www.gutenberg.org
/files/600/600-h/600-h.htm; F. Dostoyevsky, *Notes from the Underground
and the Grand Inquisitor*, trans. R. E. Matlaw (New York: Penguin, 2003); F.
Dostoyevsky, *Notes from the Underground*, trans. C. Garnett, ed. C. Guignon
and K. Aho (Indianapolis, IN: Hackett, 2009).

25. Mihajlov, "Life = Freedom."

26. C. Guignon and K. Aho, "Introduction," in Dostoyevsky, *Notes
from the Underground*, trans. Garnett, ed. Guignon and Aho.

27. D. Moon, *The Abolition of Serfdom in Russia: 1762–1907* (New York:
Routledge, 2014).

28. Sources for the insights contained in the next few paragraphs:
Guignon and Aho, "Introduction"; S. S. J. Murphy, "The Debate Around
Nihilism in 1860s Russian Literature," *Slovo* 28, no. 2 (2016): 48–68; R.
Freeborn, *The Russian Revolutionary Novel: Turgenev to Pasternak* (Cam-
bridge, UK: Cambridge University Press, 1985).

29. N. N. Taleb, *Antifragile: Things That Gain from Disorder* (New York:
Random House, 2012).

30. T. Sowell, *A Conflict of Visions: Ideological Origins of Political Struggles* (New York: Basic Books, 2007).

31. F. Dostoevsky, *Notes from the Underground*, pt. 1, chap. 8, p. 2,
Page by Page Books, accessed August 22, 2020, www.pagebypagebooks

.com/Fyodor_Dostoevsky/Notes_from_the_Underground/Part_I_Chapter
_VIII_p2.html.

32. A. Sen, "Freedom of Choice: Concept and Content," WIDER
working papers, August 22, 1987, accessed August 22, 2020, https://
ageconsearch.umn.edu/record/295553/files/WP25.pdf.

33. None of this is to say we always want freedom. As the doctor Atul
Gawande notes, patients often don't want the freedom and autonomy their
doctors offer. They want choices to be made for them. Gawande cites a personal example of his daughter being rushed to the emergency room. When
asked whether he wanted her intubated, his response was to want the choice
made for him. As he puts it, he needed his daughter's physicians "to bear
the responsibility: they could live with the consequences, good or bad." Gawande also brings survey data to bear on his claim that we don't always want
autonomy. He notes that 65 percent of people say that if they get cancer they
would want to choose their own treatment. However, it turns out that only
12 percent of people with cancer actually want to do so. Yet at least here the
choice to not make a choice is still a choice. Autonomy is something we can
give away and yet still retain. See A. Gawande, *Complications: A Surgeon's
Notes on an Imperfect Science* (New York: Profile Books, 2010).

34. J. Habermas, *Between Facts and Norms: Contributions to a Discourse
Theory of Law and Democracy*, trans. W. Rehg (Cambridge, MA: MIT Press,
1998), 306.

35. H. Mance, "Britain Has Had Enough of Experts, Says Gove,"
Financial Times, June 3, 2016, www.ft.com/content/3be49734-29cb-11e6
-83e4-abc22d5d108c.

36. J. Henrich, *The Secret of Our Success: How Culture Is Driving Human
Evolution, Domesticating Our Species, and Making Us Smarter* (Princeton, NJ:
Princeton University Press, 2017), 26, 102.

37. See, for example, Steven Pinker's lecture on Bayesian reasoning,
Spring 2019–2020: https://harvard.hosted.panopto.com/Panopto/Pages
/Viewer.aspx?id=921ab5c6-3f83-450d-b23f-ab3b0140eeae.

38. Berlin, "Four Concepts of Liberty."

39. B. Wootton, *In a World I Never Made: Autobiographical Reflections*
(London: Allen and Unwin, 1967), 279.

40. K. R. Thompson, W. A. Hochwarter, and N. J. Mathys, "Stretch
Targets: What Makes Them Effective?," *Academy of Management Perspec-
tives* 11, no. 3 (1997): 48–60; S. B. Sitkin et al., "The Paradox of Stretch
Goals: Organizations in Pursuit of the Seemingly Impossible," *Academy of
Management Review* 36 (2011): 544–566.

41. A. D. Manning, D. B. Lindenmayer, and J. Fischer, "Stretch Goals and Backcasting: Approaches for Overcoming Barriers to Large-Scale Ecological Restoration," *Restoration Ecology* 14, no. 4 (2006): 487–492.

42. S. B. Sitkin, C. C. Miller, and K. E. See, "The Stretch Goal Paradox," *Harvard Business Review*, January–February 2017, https://hbr .org/2017/01/the-stretch-goal-paradox.

43. "George Orwell's 1940 Review of *Mein Kampf*," BookMarks, orig. pub. March 21, 1940, reproduced April 26, 2017, https://bookmarks .reviews/george-orwells-1940-review-of-mein-kampf/.

44. D. Adams, *The Hitch-Hiker's Guide to the Galaxy*, pt. 11, BBC, January 24, 1980, program transcript, accessed August 22, 2020, www .clivebanks.co.uk/THHGTTG/THHGTTGradio11.htm.

45. Sitkin, Miller, and See, "The Stretch Goal Paradox."

46. D. M. Rousseau, "Organizational Behavior in the New Organizational Era," *Annual Review of Psychology* 48, no. 1 (1997): 515–546.

47. S. Sherman and S. Kerr, "Stretch Goals: The Dark Side of Asking for Miracles," CNN Money, orig. pub. November 13, 1995, accessed August 22, 2020, https://money.cnn.com/magazines/fortune/fortune _archive/1995/11/13/207680/index.htm.

48. S. Kerr and S. Landauer, "Using Stretch Goals to Promote Organizational Effectiveness and Personal Growth: General Electric and Goldman Sachs," *Academy of Management Perspectives* 18, no. 4 (2004): 134–138.

49. Sitkin, Miller, and See, "The Stretch Goal Paradox."

50. M. Gaim, S. Clegg, and M. P. E. Cunha, "Managing Impressions Rather Than Emissions: Volkswagen and the False Mastery of Paradox," *Organization Studies* (2019), https://doi.org/10.1177/0170840619891199.

51. Sitkin, Miller, and See, "The Stretch Goal Paradox."

52. Sitkin, Miller, and See, "The Stretch Goal Paradox."

53. D. Kahneman, *Thinking, Fast and Slow* (London: Penguin, 2012).

54. Sitkin, Miller, and See, "The Stretch Goal Paradox."

55. Manning, Lindenmayer, and Fischer, "Stretch Goals and Backcasting." See also "Rewilding the Scottish Highlands," Trees for Life, updated 2020, https://treesforlife.org.uk/.

56. R. Niebuhr, *Moral Man and Immoral Society: A Study in Ethics and Politics* (Louisville, KY: Westminster John Knox Press, 2013 [1932]). For a stirring modern invocation of Niebuhr, see C. Hedges, *Wages of Rebellion: The Moral Imperative of Revolt* (New York: Nation Books, 2015).

57. Niebuhr, as cited in P. Merkley, *Reinhold Niebuhr: A Political Account* (Quebec: McGill-Queen's University Press, 1975).

58. Niebuhr, *Moral Man and Immoral Society.*

6. Spite and Politics

1. M. Holmes, "Introduction: The Importance of Being Angry: Anger in Political Life," *European Journal of Social Theory* 7, no. 2 (2004): 123–132.

2. CNN, "Exit Polls," November 23, 2016, https://edition.cnn.com /election/2016/results/exit-polls.

3. A. W. Geiger, "For Many Voters, It's Not Which Presidential Candidate They're for but Which They're Against," September 2, 2016, www .pewresearch.org/fact-tank/2016/09/02/for-many-voters-its-not-which -presidential-candidate-theyre-for-but-which-theyre-against/.

4. CNN, "Exit Polls."

5. The same pattern was found in Clinton supporters, but at a lower level. Thirteen percent of people who said they would feel bad if Trump won were Trump voters, but only 9 percent of people who said they would feel bad if Clinton won were Clinton voters. Thirty-three percent of people who said they would be concerned if Trump won were Trump voters, but only 19 percent of people who said they would be concerned if Clinton won were Clinton voters. Finally, 2 percent of people who said they would be scared if Trump won were Trump voters, whereas 1 percent of people who said they would be scared if Clinton won were Clinton voters.

6. T. Boggioni, "Bernie-or-Bust Voter: At Least Trump Will Bring Change—'Even if It's Like a Nazi-Type Change,'" *Raw Story*, May 27, 2016, www.rawstory.com/2016/05/bernie-or-bust-voter-at-least-trump -will-bring-change-even-if-its-like-a-nazi-type-change/.

7. J. Noble, "Sanders Says Clinton's Agenda Matches His Own, but Backers Remain Skeptical," *Des Moines Register*, October 5, 2016, https:// eu.desmoinesregister.com/story/news/politics/2016/10/05/sanders-says -clintons-agenda-matches-his-own-but-backers-remain-skeptical/91541 064/.

8. M. B. Petersen, M. Osmundsen, and K. Arceneaux, "A 'Need for Chaos' and the Sharing of Hostile Political Rumors in Advanced Democracies," PsyArXiv Preprints, September 1, 2018, https://doi.org/10.31234/osf .io/6m4ts.

9. A point made by Zach Carter: Z. Carter, "The Democratic Party Confronts the Chaos Vote," *Huffington Post*, July 12, 2019, www.huffpost .com/entry/democratic-party-chaos-vote_n_5de95ab6e4b0913e6f8d3d5e.

10. A. Hamilton, *Federalist* no. 1, October 27, 1787, retrieved from Founders Online, National Archives, https://founders.archives.gov /documents/Hamilton/01-04-02-0152. Original source: *The Papers of Alexander Hamilton*, vol. 4, *January 1787–May 1788*, ed. Harold C. Syrett (New York: Columbia University Press, 1962), 301–306.

11. R. Di Tella and J. J. Rotemberg, "Populism and the Return of the 'Paranoid Style': Some Evidence and a Simple Model of Demand for Incompetence as Insurance Against Elite Betrayal," *Journal of Comparative Economics* 46, no. 4 (2018): 988–1005.

12. I. Bohnet and R. Zeckhauser, "Trust, Risk and Betrayal," *Journal of Economic Behavior and Organization* 55, no. 4 (2004): 467–484.

13. For the statistically minded of you, the p-value of this effect was $p=0.07$, and the authors made no correction to alpha for performing multiple statistical tests.

14. J. Graham, J. Haidt, and B. A. Nosek, "Liberals and Conservatives Rely on Different Sets of Moral Foundations," *Journal of Personality and Social Psychology* 96, no. 5 (2009): 1029–1046.

15. H. Gautney, "Dear Democratic Party: It's Time to Stop Rigging the Primaries," *The Guardian*, June 11, 2018, www.theguardian.com /commentisfree/2018/jun/11/democrat-primary-elections-need-reform.

16. Gautney, "Dear Democratic Party."

17. Office of the Director of National Intelligence, *Intelligence Community Assessment: Assessing Russian Activities and Intentions in Recent US Elections*, January 6, 2017, www.dni.gov/files/documents/ICA_2017_01 .pdf.

18. K. East, "Top DNC Staffer Apologizes for Email on Sanders' Religion," *Politico*, July 23, 2016, www.politico.com/story/2016/07/top -dnc-staffer-apologizes-for-email-on-sanders-religion-226072.

19. S. Bordo, *The Destruction of Hillary Clinton* (London: Melville House, 2017), 77.

20. A. J. Gaughan, "Was the Democratic Nomination Rigged? A Reexamination of the Clinton-Sanders Presidential Race," *University of Florida Journal of Law and Public Policy* 29 (2018): 309–358.

21. Bernie Sanders, "This is not the time for a protest vote," Facebook, September 17, 2016, www.facebook.com/berniesanders/photos /a.324119347643076/1157189241002745/?type=3.

22. W. Blitzer, "Interview with Sen. Bernie Sanders," CNN, July 6, 2016, http://transcripts.cnn.com/TRANSCRIPTS/1607/06/wolf.01 .html.

23. J. G. Voelkel and M. Feinberg, "Morally Reframed Arguments Can Affect Support for Political Candidates," *Social Psychological and Personality Science* 9, no. 8 (2018): 917–924.

24. R. Reich, "Why You Must Vote for Hillary," *Newsweek*, September 19, 2016, www.newsweek.com/robert-reich-why-you-must-vote -hillary-500197.

25. H. Parry, F. Chambers, and K. Rahman, "Lady Gaga Stages Protest Outside Trump Tower While Demi Lovato Is Forced to Apologize for Tweet Mocking the Donald as Hillary's Hollywood Stars Struggle to Cope," *Daily Mail*, November 9, 2016, www.dailymail.co.uk/news/article -3918926/Hollywood-starts-panic-results-aren-t-going-Clinton-s-way .html.

26. Bordo, *The Destruction of Hillary Clinton*, 67. Albright had been voicing this opinion for many years previously: M. O'Connor, "A Brief History of Taylor Swift's 'Special Place in Hell for Women,'" *The Cut*, March 6, 2013, www.thecut.com/2013/03/brief-history-of-taylor-swifts-hell-quote .html.

27. J. D. Williams, "Sanders Supporters: It's Infuriating to Be Told We Have to Vote for Hillary. But We Do," *The Nation*, November 4, 2016, www.thenation.com/article/sanders-supporters-its-infuriating-to-be-told -we-have-to-vote-for-hillary-but-we-do/.

28. Data from the Cooperative Congressional Election Study, which surveyed around fifty thousand people, as reported by D. Kurtzleben, "Here's How Many Bernie Sanders Supporters Ultimately Voted for Trump," *NPR*, August 24, 2017, www.npr.org/2017/08/24/545812242/1 -in-10-sanders-primary-voters-ended-up-supporting-trump-survey -finds?t=1586332242609.

29. Nearly half of Sanders supporters who voted for Trump believed that whites were not actually privileged in America; see Kurtzleben, "Here's How Many Bernie Sanders Supporters Ultimately Voted for Trump."

30. M. Krieger, "'Bernie or Bust': Over 50,000 Sanders Supporters Pledge to Never Vote for Hillary," *Liberty Blitzkrieg*, March 3, 2016, https://libertyblitzkrieg.com/2016/03/03/bernie-or-bust-over-50000 -sanders-supporters-pledge-to-never-vote-for-hillary/.

31. L. Gambino, "Trump: Hillary Clinton May Be 'Most Corrupt Person Ever' to Run for President," *The Guardian*, June 22, 2016, www .theguardian.com/us-news/2016/jun/22/donald-trump-hillary-clinton -corrupt-person-president.

32. See E. F. Bloomfield and G. Tscholl, "Analyzing Warrants and Worldviews in the Rhetoric of Donald Trump and Hillary Clinton: Burke and Argumentation in the 2016 Presidential Election," *Journal of the Kenneth Burke Society* 13, no. 2 (2018), https://digitalscholar ship.unlv.edu/cgi/viewcontent.cgi?article=1036&context=comm_fac _articles.

33. Di Tella and Rotemberg, "Populism and the Return of the 'Paranoid Style.'"

34. R. Berman, "Who Will Grab the Bernie-or-Bust and the Never-Trump Vote?," *The Atlantic*, June 9, 2016, www.theatlantic.com/politics /archive/2016/06/who-will-grab-the-bernie-or-bust-and-the-never-trump -vote/486254/.

35. D. Roberts, "Latest WikiLeaks Dump Ties Clinton Foundation to Personal Enrichment Claims," *The Guardian*, October 27, 2016, www.theguardian.com/world/2016/oct/27/wikileaks-bill-clinton -foundation-emails.

36. Donald J. Trump (@realDonaldTrump), "The system is rigged," Twitter, July 5, 2016, 8:37 a.m., https://twitter.com/realDonaldTrump /status/750352884106223616.

37. Bordo, *The Destruction of Hillary Clinton*, 65.

38. E. Crockett, "Obama: 'Not Me, Not Bill, Nobody' Has Been More Qualified for President Than Hillary," *Vox*, June 27, 2016, www .vox.com/2016/7/27/12306702/democratic-convention-obama-hillary -clinton-bill-qualified.

39. Crockett, "Obama: 'Not Me, Not Bill, Nobody' Has Been More Qualified."

40. S. Clench, "The Moment Still Haunting Hillary Clinton 24 Years Later," *News.com.au*, September 5, 2016, www.news.com.au/finance /work/leaders/the-moment-still-haunting-hillary-clinton-24-years-later /news-story/57ee06c9ca156a82339802987a938380.

41. K. Reilly, "Beyoncé Reclaims Hillary Clinton's 'Baked Cookies' Comment at Rally," *Time*, November 5, 2016, https://time.com/4559565 /hillary-clinton-beyonce-cookies-teas-comment/.

42. Bordo, *The Destruction of Hillary Clinton*, 45.

43. Associated Press, "WikiLeaks: Clinton Avoided Criticism of Wall Street in Goldman Sachs Speeches," *The Guardian*, October 16, 2016, www.theguardian.com/us-news/2016/oct/16/wikileaks -hillary-clinton-wall-street-goldman-sachs-speeches.

44. H. R. Clinton, *What Happened* (New York: Simon and Schuster, 2017), 413.

45. B. Weiss, "Jonathan Haidt on the Cultural Roots of Campus Rage," *Wall Street Journal*, April 14, 2017, www.wsj.com/articles /jonathan-haidt-on-the-cultural-roots-of-campus-rage-1491000676.

46. Bordo, *The Destruction of Hillary Clinton*, 49.

47. J. A. Minson and B. Monin, "Do-Gooder Derogation: Disparaging Morally Motivated Minorities to Defuse Anticipated Reproach," *Social Psychological and Personality Science* 3, no. 2 (2012): 200–207.

48. L. Guiso et al., *Demand and Supply of Populism* (London: Centre for Economic Policy Research, 2017).

49. M. Crockett, "Why Your Brain Loves Feeling Outraged and Punishing People's Bad Behavior," Big Think, October 9, 2017, www.youtube.com/watch?v=3z3UoO8JdOo.

50. N. Loiseau, "Brexit Is a 'Lose-Lose Situation' for UK and EU, Says Minister," French Embassy in London, October 21, 2018, https://uk.ambafrance.org/Brexit-Ball-is-in-UK-s-court-on-Irish-border-issue-says-Minister.

51. R. Taylor, "Cameron Refuses to Apologise to UKIP," *The Guardian*, April 4, 2006, www.theguardian.com/politics/2006/apr/04/conservatives.uk.

52. R. Mason and A. Asthana, "David Cameron: Leave Vote Would Be Economic Bomb for UK," *The Guardian*, June 6, 2016, www.theguardian.com/politics/2016/jun/06/david-cameron-brexit-would-detonate-bomb-under-uk-economy.

53. T. Ross, "Nigel Farage: Migrants Could Pose Sex Attack Threat to Britain," *The Telegraph*, June 4, 2016, www.telegraph.co.uk/news/2016/06/04/nigel-farage-migrants-could-pose-sex-attack-threat-to-britain/.

54. M. Dathan, "'Brexit Campaigners Are EXTREMISTS!' Labour's EU Leader Alan Johnson Launches Extraordinary Attack on Vote Leave as Corbyn Launches the Party's Red Battle Bus," *Daily Mail*, May 10, 2016, www.dailymail.co.uk/news/article-3582763/We-need-tent-protect-storm-soaked-embattled-Jeremy-Corbyn-mocks-Labour-critics-poor-election-results-launches-party-s-pro-EU-battle-bus-pouring-rain.html.

55. J. Daley, "Why Am I Considered a Bigot or an Idiot for Wanting Britain to Leave the EU?," *The Telegraph*, March 5, 2016, www.telegraph.co.uk/opinion/2016/03/17/why-am-i-considered-a-bigot-or-an-idiot-for-wanting-britain-to-l/.

56. P. Dallison, "Farage on Remainers: They Think 'We're Stupid, We're Ignorant, We're Racist,'" *Politico*, May 22, 2019, www.politico.eu/article/nigel-farage-on-brexit-remainers-they-think-we-are-thick-stupid-ignorant-racist-european-parliament-election-uk/.

57. T. Wallace, "Voters Really Have Had Enough of Experts: Trust in Economists Has Slumped Since Referendum," *The Telegraph*, November 22, 2019, www.telegraph.co.uk/business/2019/11/22/voters-really-have-had-enough-experts-trust-economists-has-slumped/.

58. "YouGov/Times Survey Results," YouGov, 2016, https://d25d2506sfb94s.cloudfront.net/cumulus_uploads/document/atmwrgevvj/TimesResults_160622_EVEOFPOLL.pdf.

59. P. Turchin, *Historical Dynamics: Why States Rise and Fall* (Princeton, NJ: Princeton University Press, 2003).

60. J. C. Fournier et al., "Antidepressant Drug Effects and Depression Severity: A Patient-Level Meta-analysis," *JAMA* 303, 1 (2010): 47–53. Obviously, if you are taking antidepressants and decide to stop using them, it is something you should do in conjunction with a medical professional.

61. J. Read, C. Cartwright, and K. Gibson, "Adverse Emotional and Interpersonal Effects Reported by 1829 New Zealanders While Taking Antidepressants," *Psychiatry Research* 216, no. 1 (2014): 67–73.

62. L. A. Pratt, D. J. Brody, Q. Gu, "Antidepressant Use Among Persons Aged 12 and Over: United States, 2011–2014," NCHS Data Brief no. 283, August 2017, www.cdc.gov/nchs/data/databriefs/db283.pdf.

63. G. Iacobucci, "NHS Prescribed Record Number of Antidepressants Last Year," *BMJ* 364 (2019): l1508.

64. K. Daniels and J. C. Abma, "Current Contraceptive Status Among Women Aged 15–49: United States, 2015–2017," NCHS Data Brief no. 327, December 2018, www.cdc.gov/nchs/data/databriefs/db327-h.pdf.

65. J. P. Del Río et al., "Steroid Hormones and Their Action in Women's Brains: The Importance of Hormonal Balance," *Frontiers in Public Health* 6 (2018): 141.

7. Spite and the Sacred

1. Romans 12:19 (King James Version).

2. Specifically Ezekiel 25:17 (King James Version).

3. C. Hackett and D. McClendon, "Christians Remain World's Largest Religious Group, but They Are Declining in Europe," Pew Research Center, April 5, 2017, www.pewresearch.org/fact-tank/2017/04/05/christians-remain-worlds-largest-religious-group-but-they-are-declining-in-europe/.

4. K. Laurin, "Belief in God: A Cultural Adaptation with Important Side Effects," *Current Directions in Psychological Science* 26, no. 5 (2017): 458–463; K. Laurin et al., "Outsourcing Punishment to God: Beliefs in Divine Control Reduce Earthly Punishment," *Proceedings of the Royal Society B: Biological Sciences* 279, no. 1741 (2012): 3272–3281.

5. Laurin, "Belief in God"; Laurin et al., "Outsourcing Punishment to God."

6. Laurin et al., "Outsourcing Punishment to God."

7. Laurin et al., "Outsourcing Punishment to God."

8. F. Nietzsche, *On the Genealogy of Morality*, ed. K. Ansell-Pearson, trans. C. Diethe (Cambridge, UK: Cambridge University Press, 2006 [1887]), 20.

9. Z. Reeve, "Terrorism as Parochial Altruism: Experimental Evidence," *Terrorism and Political Violence* (2019): 1–24.

10. C. McCauley, "How Many Suicide Terrorists Are Suicidal?," *Behavioral and Brain Sciences* 37, no. 4 (2014): 373–374.

11. M. Eswaran and H. M. Neary, "Decentralized Terrorism and Social Identity" (working paper, Microeconomics.ca, 2018) https://ideas.repec.org/p/ubc/pmicro/tina_marandola-2018-4.html.

12. G. LaFree and L. Dugan, "How Does Studying Terrorism Compare to Studying Crime?," in *Terrorism and Counter-Terrorism*, vol. 5, *Sociology of Crime, Law and Deviance*, ed. M. Deflem (Bingley, UK: Emerald Group Publishing, 2004), 53–74.

13. McCauley, "How Many Suicide Terrorists Are Suicidal?"

14. H. N. Qirko, "Altruism in Suicide Terror Organizations," *Zygon* 44, no. 2 (2009): 289–322; S. Atran, *Talking to the Enemy: Faith, Brotherhood and the (Un)Making of Terrorists* (New York: HarperCollins, 2010); S. Atran, "The Devoted Actor: Unconditional Commitment and Intractable Conflict Across Cultures," *Current Anthropology* 57, no. S13 (2016): S192–S203.

15. K. Jacques and P. J. Taylor, "Male and Female Suicide Bombers: Different Sexes, Different Reasons?," *Studies in Conflict and Terrorism* 31, no. 4 (2008): 304–326.

16. A. D. Blackwell, "Middle-Class Martyrs: Modeling the Inclusive Fitness Outcomes of Palestinian Suicide Attack," ResearchGate, January 2008, www.researchgate.net/publication/323883656_Middle-class_martyrs_Modeling_the_inclusive_fitness_outcomes_of_Palestinian_suicide_attack.

17. Jacques and Taylor, "Male and Female Suicide Bombers."

18. Eswaran and Neary, "Decentralized Terrorism and Social Identity."

19. J. Stern, "Beneath Bombast and Bombs, a Caldron of Humiliation," *Los Angeles Times*, June 6, 2004.

20. T. H. Kean et al., *The 9/11 Commission Report: Final Report of the National Commission on Terrorist Attacks upon the United States*, 2004, www.9-11commission.gov/report/911Report.pdf.

21. *U.S. Government Efforts to Counter Violent Extremism: Hearing Before the Subcommittee on Emerging Threats and Capabilities of the Committee on Armed Services*, US Senate, 111th Cong., Second Session (March 10,

2010) (statement of S. Atran), www.govinfo.gov/content/pkg/CHRG
-111shrg63687/html/CHRG-111shrg63687.htm.

22. A. Speckhard and K. Akhmedova, "Black Widows: The Chechen
Female Suicide Terrorists," in *Female Suicide Terrorists*, ed. Y. Schweitzer
(Tel Aviv: Jaffee Center, 2006).

23. Speckhard and Akhmedova, "Black Widows."

24. S. Atran, "Genesis of Suicide Terrorism," *Science* 299, no. 5612
(2003): 1534–1539.

25. S. Atran and H. Sheikh, "Dangerous Terrorists as Devoted Actors," in *Evolutionary Psychology: Evolutionary Perspectives on Social Psychology*, ed. V. Zeigler-Hill, L. L. M. Welling, and T. K. Shackelford (Cham,
Switzerland: Springer, 2015), 401–416.

26. Atran, "Genesis of Suicide Terrorism."

27. J. Ginges and S. Atran, "War as a Moral Imperative: Not Practical
Politics by Other Means," *Proceedings of the Royal Society B: Biological Sciences*
27 (2011): 2930–2938.

28. S. Atran, "Why People Become Terrorists," Full Circle, October
26, 2017, www.youtube.com/watch?v=7SFc1l62FJ4.

29. N. Hamid et al., "Neuroimaging 'Will to Fight' for Sacred Values:
An Empirical Case Study with Supporters of an Al Qaeda Associate," *Royal
Society Open Science* 6, no. 6 (2019): 181585.

30. C. Pretus et al., "Ventromedial and Dorsolateral Prefrontal Interactions Underlie Will to Fight and Die for a Cause," *Social Cognitive and
Affective Neuroscience* 14, no. 6 (2019): 569–577.

31. Hamid et al., "Neuroimaging 'Will to Fight' for Sacred Values."

32. J. Ginges et al., "Sacred Bounds on Rational Resolution of Violent
Political Conflict," *Proceedings of the National Academy of Sciences* 104, no. 18
(2007): 7357–7360.

33. C. Pretus et al., "Neural and Behavioral Correlates of Sacred Values
and Vulnerability to Violent Extremism," *Frontiers in Psychology* 9 (2018):
2462.

34. S. Lyons-Padilla et al., "Belonging Nowhere: Marginalization and
Radicalization Risk Among Muslim Immigrants," *Behavioral Science and
Policy* 1, no. 2 (2015): 1–12.

35. G. S. Berns et al., "The Price of Your Soul: Neural Evidence for the
Non-utilitarian Representation of Sacred Values," *Philosophical Transactions
of the Royal Society B: Biological Sciences* 367, no. 1589 (2012): 754–762.

36. M. J. Souza, S. E. Donohue, and S. A. Bunge, "Controlled Re-
trieval and Selection of Action-Relevant Knowledge Mediated by Partially

Overlapping Regions in Left Ventrolateral Prefrontal Cortex," *Neuroimage* 46, no. 1 (2009): 299–307.

37. Reeve, "Terrorism as Parochial Altruism."

38. A. Pedahzur, "Toward an Analytical Model of Suicide Terrorism: A Comment," *Terrorism and Political Violence* 16 (2004): 841–844.

39. Speckhard and Akhmedova, "Black Widows."

40. R. Janoff-Bulman, *Shattered Assumptions: Towards a New Psychology of Trauma* (New York: Free Press, 1992). And, yes, the following seems to be based on a very privileged view of the world.

41. Speckhard and Akhmedova, "Black Widows."

42. J. Reuter, *Chechnya's Suicide Bombers: Desperate, Devout, or Deceived?*, American Committee for Peace in Chechnya, August 23, 2004, https://jamestown.org/wp-content/uploads/2011/01/Chechen_Report _FULL_01.pdf?x17103.

43. R. A. Pape, *Dying to Win: The Strategic Logic of Suicide Terrorism* (New York: Random House, 2006); Qirko, "Altruism in Suicide Terror Organizations."

44. H. Sheikh, J. Ginges, and S. Atran, "Sacred Values in the Israeli-Palestinian Conflict: Resistance to Social Influence, Temporal Discounting and Exit Strategies," *Annals of the New York Academy of Sciences* 1299 (2013): 11–24.

45. Reeve, "Terrorism as Parochial Altruism."

46. F. Pratto et al., "Social Dominance Orientation: A Personality Variable Predicting Social and Political Attitudes," *Journal of Personality and Social Psychology* 67, no. 4 (1994): 741–763.

47. Pratto et al., "Social Dominance Orientation"; A. F. Lemieux and V. H. Asal, "Grievance, Social Dominance Orientation, and Authoritarianism in the Choice and Justification of Terror Versus Protest," *Dynamics of Asymmetric Conflict* 3, no. 3 (2010): 194–207.

48. H. M. Crowson and J. A. Brandes, "Differentiating Between Donald Trump and Hillary Clinton Voters Using Facets of Right-Wing Authoritarianism and Social-Dominance Orientation: A Brief Report," *Psychological Reports* 120, no. 3 (2017): 364–373; B. L. Choma and Y. Hanoch, "Cognitive Ability and Authoritarianism: Understanding Support for Trump and Clinton," *Personality and Individual Differences* 106 (2017): 287–291. If you're already annoyed at this paragraph, do not read what the Choma and Hanoch paper found. Seriously.

49. W. B. Swann Jr. et al., "When Group Membership Gets Personal: A Theory of Identity Fusion," *Psychological Review* 119, no. 3 (2012):

441–456; S. Atran, "Measures of Devotion to ISIS and Other Fighting and Radicalized Groups," *Current Opinion in Psychology* 35 (2020): 103–107.

50. W. B. Swann Jr. et al., "What Makes a Group Worth Dying For? Identity Fusion Fosters Perception of Familial Ties, Promoting Self-Sacrifice," *Journal of Personality and Social Psychology* 106, no. 6 (2014): 912–926.

51. H. Sheikh, Á. Gómez, and S. Atran, "Empirical Evidence for the Devoted Actor Model," *Current Anthropology* 57, no. S13 (2016): S204–209.

52. H. Whitehouse and J. A. Lanman, "The Ties That Bind Us: Ritual, Fusion and Identification," *Current Anthropology* 55, no. 6 (2014): 674–695.

53. H. Whitehouse et al., "The Evolution of Extreme Cooperation via Shared Dysphoric Experiences," *Scientific Reports* 7 (2017): 44292.

54. Whitehouse et al., "The Evolution of Extreme Cooperation via Shared Dysphoric Experiences."

55. See H. Whitehouse, "Dying for the Group: Towards a General Theory of Extreme Self-Sacrifice," *Behavioral and Brain Sciences* 41 (2018): e192.

56. Swann et al., "What Makes a Group Worth Dying For?"

57. H. Whitehouse et al., "Brothers in Arms: Libyan Revolutionaries Bond like Family," *Proceedings of the National Academy of Sciences* 111, no. 50 (2014): 17783–17785.

58. Atran, "The Devoted Actor."

59. M. Sageman, "The Stagnation in Terrorism Research," *Terrorism and Political Violence* 26, no. 4 (2014): 565–580; *U.S. Government Efforts to Counter Violent Extremism*, statement of S. Atran.

60. E. Benmelech, C. Berrebi, and E. F. Klor, "Counter-Suicide-Terrorism: Evidence from House Demolitions," *Journal of Politics* 77, no. 1 (2015): 27–43. The effects of this decrease over time. Furthermore, precautionary house demolitions (ones conducted on the basis of where a house is located, but not related to the owner or their activities) cause a significant increase in the number of suicide terror attacks.

61. S. Atran, "The Moral Logic and Growth of Suicide Terrorism," *Washington Quarterly* 29, no. 2 (2006): 127–147.

62. Ginges et al., "Sacred Bounds on Rational Resolution of Violent Political Conflict."

63. Kean et al., *The 9/11 Commission Report*.

64. O. P. Hauser et al., "Cooperating with the Future," *Nature* 511, no. 7508 (2014): 220–223.

8. The Future of Spite

1. D. W. Lovell, "The Concept of the Proletariat in the Work of Karl Marx" (PhD diss., Australian National University, 1984), reproduced at https://openresearch-repository.anu.edu.au/bitstream/1885/124605/2/b12112537_Lovell_David_W.pdf.

2. J. Ronson, "How One Stupid Tweet Blew Up Justine Sacco's Life," *New York Times*, February 15, 2015; see also J. Ronson, *So You've Been Publicly Shamed* (New York: Riverhead Books, 2016).

3. D. Murray, *The Madness of Crowds: Gender, Race and Identity* (London: Bloomsbury, 2019), Kindle ed.

4. *Testimony Given to a Joint Hearing Before the Subcommittee on Healthcare, Benefits, and Administrative Rules and the Subcommittee on Intergovernmental Affairs of the Committee on Oversight and Governmental Reform*, US House of Representatives, 115th Cong., Second Session (May 22, 2018) (testimony of B. Weinstein), www.govinfo.gov/content/pkg/CHRG-115hhrg32667/html/CHRG-115hhrg32667.htm.

5. P. McCormack, "What Happened at Evergreen College?," *Defiance*, October 13, 2019, www.defiance.news/def007-bret-weinstein.

6. N. A. Robins and A. Jones, eds., *Genocides by the Oppressed: Subaltern Genocide in Theory and Practice* (Bloomington: Indiana University Press, 2009).

7. As Paul Russell has noted: P. Russell, "Vice Dressed as Virtue," *Aeon*, May 22, 2020, https://aeon.co/essays/how-the-cruel-moraliser-uses-a-halo-to-disguise-his-horns.

8. C. D. Batson et al., "In a Very Different Voice: Unmasking Moral Hypocrisy," *Journal of Personality and Social Psychology* 72, no. 6 (1997): 1335–1348; C. D. Batson and E. C. Collins, "Moral Hypocrisy: A Self-Enhancement/Self-Protection Motive in the Moral Domain," in *The Handbook of Self-Enhancement and Self-Protection*, ed. M. D. Alicke and C. Sedikides (New York: Guilford, 2011), 92–111.

9. See C. D. Batson, "What's Wrong with Morality?," *Emotion Review* 3, no. 3 (2011): 230–236.

10. Batson, "What's Wrong with Morality?"

11. D. K. Marcus et al., "The Psychology of Spite and the Measurement of Spitefulness," *Psychological Assessment* 26, no. 2 (2014): 563–574.

12. M. Batey and A. Furnham, "Creativity, Intelligence, and Personality: A Critical Review of the Scattered Literature," *Genetic, Social, and General Psychology Monographs* 132 (2006): 355–429.

13. Though not greater creativity in realms such as dance, painting, and poetry. See C. D. Davis, J. C. Kaufman, and F. H. McClure, "Non-cognitive

Constructs and Self-Reported Creativity by Domain," *Journal of Creative Behavior* 45, no. 3 (2011): 188–202.

14. B. Bakker, G. Schumacher, and M. Rooduijn, "The Populist Appeal: Personality and Anti-establishment Communication," *Journal of Politics* (forthcoming), accessed August 2020, https://psyarxiv.com/n3je2/download?format=pdf.

15. See S. Johnson, "We Need Disagreeable People to Fix Our Dishonest Institutions," Big Think, January 31, 2019, https://bigthink.com/culture-religion/eric-weinstein-intellectual-dark-web.

16. To be clear, the category of the prize is actually called "Physiology or Medicine." It is not that I am unsure which he won.

17. A. Rutherford, "He May Have Unravelled DNA, but James Watson Deserves to Be Shunned," *The Guardian*, December 1, 2014, www.theguardian.com/commentisfree/2014/dec/01/dna-james-watson-scientist-selling-nobel-prize-medal.

18. S. T. Hunter and L. Cushenbery, "Is Being a Jerk Necessary for Originality? Examining the Role of Disagreeableness in the Sharing and Utilization of Original Ideas," *Journal of Business and Psychology* 30, no. 4 (2015): 621–639.

19. C. Starmans, M. Sheskin, and P. Bloom, "Why People Prefer Unequal Societies," *Nature Human Behavior* 1, no. 4 (2017): 0082.

20. CIA, "Timeless Tips for 'Simple Sabotage,'" July 12, 2012, www.cia.gov/news-information/featured-story-archive/2012-featured-story-archive/simple-sabotage.html.

21. G. Grolleau, N. Mzoughi, and A. Sutan, "The Impact of Envy-Related Behaviors on Development," *Journal of Economic Issues* 43, no. 3 (2009): 795–808.

22. R. Chetty et al., "The Fading American Dream: Trends in Absolute Income Mobility Since 1940," *Science* 356, no. 6336 (2017): 398–406.

23. M. Boltz, K. Marazyan, and P. Villar, "Income Hiding and Informal Redistribution: A Lab-in-the-Field Experiment in Senegal," *Journal of Development Economics* 137 (2019): 78–92.

24. M. I. Norton, "Unequality: Who Gets What and Why It Matters," *Policy Insights from the Behavioral and Brain Sciences* 1, no. 1 (2014): 151–155.

25. E. Luce, *The Retreat of Western Liberalism* (London: Abacus, 2017), 123.

26. E. Javers, "Inside Obama's Bank CEOs Meeting," *Politico*, March 4, 2019, www.politico.com/story/2009/04/inside-obamas-bank-ceos-meeting-020871.

27. G. Pettigrove and K. Tanaka, "Anger and Moral Judgment," *Australasian Journal of Philosophy* 92, no. 2 (2014): 269–286.

28. B. Dubreuil, "Anger and Morality," *Topoi* 34, no. 2 (2015): 475–482.

29. M. van't Wout, L. J. Chang, and A. G. Sanfey, "The Influence of Emotion Regulation on Social Interactive Decision-Making," *Emotion* 10 (2010): 815–821.

30. The original paper is D. P. Calvillo and J. N. Burgeno, "Cognitive Reflection Predicts the Acceptance of Unfair Ultimatum Game Offers," *Judgment and Decision Making* 10, no. 4 (2015): 332–341, and the questions were retrieved from http://journal.sjdm.org/14/14715/stimuli.pdf.

31. Your intuition probably says yes. But reason tells us that, no, we can't conclude that roses are flowers. All flowers may have petals, but this does not imply that everything that has petals is necessarily a flower. If the first statement had been "Everything with petals is a flower," then we could have concluded that roses having petals meant they were flowers.

32. Intuition probably makes you think thirty. But if Jerry has the fifteenth-highest mark, this means there are fourteen people above him. If he has the fifteenth-lowest, there are also fourteen people below him. So you have fourteen people above him, fourteen below, and Jerry himself: 14 + 14 + 1 = 29 people.

33. If it takes two nurses two minutes to measure the blood pressure of two patients, then this means that one nurse takes two minutes to take the blood pressure of one patient. If you had two hundred nurses working at the same time, then, in two minutes, each of them would have taken the blood pressure of one patient, meaning that two hundred patients would have been completed. So the answer is two minutes.

34. Calvillo and Burgeno, "Cognitive Reflection Predicts the Acceptance of Unfair Ultimatum Game Offers."

35. U. Kirk, J. Downar, and P. R. Montague, "Interoception Drives Increased Rational Decision-Making in Meditators Playing the Ultimatum Game," *Frontiers in Neuroscience* 5 (2011): 49.

36. Kirk, Downar, and Montague, "Interoception Drives Increased Rational Decision-Making in Meditators Playing the Ultimatum Game."

37. B. P. Chapman et al., "Emotion Suppression and Mortality Risk over a 12-Year Follow-Up," *Journal of Psychosomatic Research* 75, no. 4 (2013): 381–385; S. Greer and T. Morris, "Psychological Attributes of Women Who Develop Breast Cancer: A Controlled Study," *Journal of Psychosomatic Research* 19, no. 2 (1975): 147–153; S. P. Thomas et al., "Anger and Cancer: An Analysis of the Linkages," *Cancer Nursing* 23, no. 5 (2000): 344–349;

M. McKenna et al., "Psychosocial Factors and the Development of Breast Cancer: A Meta-analysis," *Health Psychology* 18, no. 5 (1999): 520–531; L. D. Cameron and N. C. Overall, "Suppression and Expression as Distinct Emotion-Regulation Processes in Daily Interactions: Longitudinal and Meta-analyses," *Emotion* 18, no. 4 (2018): 465–480.

38. F. Bolle, J. H. W. Tan, and D. J. Zizzo (2010). *Vendettas*. CeDEx Discussion Paper Series, Discussion Paper No. 2010-02, www.nottingham.ac.uk/cedex/documents/papers/2010-02.pdf.

39. Dubreuil, "Anger and Morality."

40. M. E. McCullough, R. Kurzban, and B. A. Tabak, "Cognitive Systems for Revenge and Forgiveness," *Behavioral and Brain Sciences* 36, no. 1 (2013): 1–15.

41. F. Nietzsche, *On the Genealogy of Morality*, ed. K. Ansell-Pearson, trans. C. Diethe (Cambridge, UK: Cambridge University Press, 2006 [1887]).

42. McCullough, Kurzban, and Tabak, "Cognitive Systems for Revenge and Forgiveness"; M. McCullough, *Beyond Revenge: The Evolution of the Forgiveness Instinct* (New York: John Wiley and Sons, 2008).

43. P. Barclay, "Pathways to Abnormal Revenge and Forgiveness," *Behavioral and Brain Sciences* 36, no. 1 (2013): 17–18.

44. D. Kahneman, J. L. Knetsch, and R. H. Thaler, "Fairness and the Assumptions of Economics," *Journal of Business* 59, no. 4 (1986): S285–S300.

45. S. Mowe, "How to Respond to Social Injustice: An Interview with Buddhist Scholar John Makransky," *Tricycle*, Summer 2012, https://tricycle.org/magazine/arent-we-right-be-angry/.

INDEX

Aché people (Paraguay), 23, 53

Adams, Douglas, 128–129

African Americans: hierarchy-legitimizing myths, 180

aggressive dominance, 44–45

agricultural societies, 46–47, 164–165

Akhmedova, Khapta, 170, 177–178

Al Qaeda, 172

Albright, Madeleine, 146

alcohol consumption: effect on Ultimatum Game behavior, 24

all-or-nothing phenomenon, 16–17

altruism
addressing environmental issues, 185–187
costly punishment, 50
defining spite, 2–3
effects of natural selection, 94–96
identity fusion, 182
motivating suicide bombers, 166–167, 179–181
parochial, 179–181
refraining from punishing unfairness, 68

American Dream, 200–201

Ames, Mark, 33–34

anger
Braveheart Effect, 114–115
controlling, 57–58, 202–204
costly punishment, 60–61
information generated from, 59
long-term purpose of, 66
moral outrage, 56–57, 197–198
response to unfairness, 56–58, 197–198
spite voting in 2016, 147
use of costly punishment, 104–105

anonymity
fear of retaliation for spite, 84–85
gossip, 70
lack of accountability for spite, 76
online, 193–194, 196
third-party punishment, 65–66

anonymous betting games, 54

anthropology: dominance-seeking and egalitarian tendencies, 39–41

antidepressants, 89, 161–162
antisocial behavior: Ultimatum
    Game, 49
ants: Hamiltonian spite in
    reproduction, 97–98
apocalyptic residual, 36, 191
Aristotle, 26, 205
Atran, Scott, 169–172, 177,
    182–183, 185
auctions: study in behavior, 15–17
automobile manufacturers, 129–130
autonomous individual concept,
    112–113

Baader, Andreas, 11–12, 20
Baader-Meinhof group, 9–13, 176,
    191–193
Balafoutas, Loukas, 90
Barclay, Pat, 206
Bartha, Nicholas, 26
"basket of deplorables" comment,
    34, 151–152, 154
Bath, Douglas, 63
Batson, C. Daniel, 197–198
Baumeister, Roy, 114
benefits of spiteful behavior, 6
    advances made by disagreeable
        people, 198–199
    competitiveness, 90
    costly punishment, 66
    creativity, 125, 198
    existential spite, 125–126
    generosity, 37
    genetic effects on behavior,
        93–98
    Hamiltonian spite, 97
    the how and why of spite, 190
    long-term social effects of spite,
        191
    measuring harm and benefits, 4

parochial altruism, 180
psychological spite, 100
return-benefit spite, 99
social media rewarding spite,
    193–195
Ultimatum Game, 17–18
Ben-Israel, Isaac, 184
Benzodiazepines, 57
Berkshire Hathaway, 31–32
Berlin, Isaiah, 125–126
betrayal, 141–142, 145
Beyoncé, 150–151
Bible: chimerical qualities of human
    nature, 6–7
biological markets theory, 74–75
black ball, 35–36
Blackwell, Aaron, 168
Blitzer, Wolf, 145
Boehm, Christopher, 39–41
bonobos
    counterdominant behaviors, 42
    lack of spiteful behavior, 20
Borges, Jorge Luis, 103
Bostrom, Nick, 35, 191
Bowles, Sam, 50
Braveheart Effect, 114–117, 126,
    132, 146
Brexit referendum, 123, 154–159
Bryson, Bill, 63
Bshary, Redouan, 100
Buddhist meditations, 204–205, 207
Buffett, Warren, 31–32
Burgeno, Jessica, 203
Bush, George W., 34
business
    accusations of Clinton's
        involvement with big
        business, 148–149
    incentivizing profit-maximizing
        businesses, 206–207

spiteful acts between business rivals, 30–32
stretch goals, 126–127, 129–133

call-out culture, 194–196
Calvillo, Dustin, 203
Cameron, David, 154–155
cancel culture, 194–195
capitalism, 127–128, 192, 195
Carney, Mark, 155
Chaldean community, 69–70
character of spiteful people, 36–37
cheap spite, 69–71
cheating: testing free will, 114
Chechen Black Widows, 170–171, 177
Chechen separatists, 178
Chen, Xiaojie, 104
Cheney, Dick, 172
Chernyshevsky, Nikolai, 118–119
children
    infanticide, 27–30
    reactance, 113
    as weapons in spousal spite, 27
chimpanzees
    costly punishment, 64
    counterdominant behaviors, 42
    lack of spiteful behavior, 20–21
    money-burning games, 54–55
Christianity as a dominance mechanism, 165–166
Clinton, Hillary, 136–139, 141–153
cognitive reflection, 203–204
communistic societies, 39
compassion, 206–207
competition
    disagreeableness associated with creativity, 198–199
    dominant spite and competition for status, 80–82

effect of serotonin on dominant spite, 88–91
evolution of psychological spite, 100–101
fair-minded people punishing unfairness, 76
Hamiltonian spite targeting competitors, 96–99
increasing performance in spiteful individuals, 90–91, 198–199
labor oversupply, 160
long-term role of spite, 190
motivating dominant spite, 86–87, 135–136
political competition and dominant spite, 135–136
purpose of costly punishment, 105–110
competitive behavior, 81
A Conflict of Visions (Sowell), 119–120
conformity
    execution hypothesis, 44
    tragedy of the commons, 50–51
Constrained Vision, 119–120, 124
contraception, 162
controlling anger, 57–58, 202–204
cooperation
    as benefit of spite, 6
    counterdominant and dominant spite, 83
    defining spite, 2–3, 5
    Dictator Game, 48–49
    effects of natural selection, 94–96
    games for boosting, 185–187
    gods enforcing, 164–165
    long-term social effects of spite, 191

cooperation (*continued*)
  nonfinancial punishment
    encouraging, 71
  purpose of costly punishment,
    105–110
  reciprocity, 49–50
  social rewards, 190
  spite improving, 66
  tragedy of the commons, 50–51
  trust and, 59–60
  Ultimate Game Behavior
    influencing, 49
cooperators, 104
cost-benefit analysis
  four basic behavioral
    interactions, 2
  genetic effects on behavior,
    94–96
  motivations of suicide bombers,
    172–174
  selfishness and empathy, 58–60
costly punishment
  Baader-Meinhof group, 192–193
  cheap spite as alternative to,
    69–70
  do-gooder derogation, 72–75
  online spite, 194
  Parrondo's paradox, 104
  the price of justice, 55–56
  purpose of, 105–110
  the role of emotions in, 104–105
  societal benefits of, 62–63
  suicide bombing, 176
  third-party, 64–66, 192–194,
    197
  tragedy of the commons, 50–51
  translating anger into, 60–61
  2016 election, 137–138,
    142–144
  See also punishment

costly signaling, 68
costs of spiteful behavior
  auction study, 15–17
  Brexit, 158–159
  effect of antidepressants on spite,
    161–162
  genetic effects, 94–98
  the how and why of spite, 190
  parochial altruism, 179–180
  social media reducing, 193–195
  suicide bombers, 171–172
counterdominant behaviors
  cheap spite, 69–70
  factors influencing, 46–47
  guilt and selfless behavior,
    51–53
  in humans and primates, 41–43
  long-term social effects of spite,
    191
  punishing generosity, 74
  punishing unfairness, 75–76
  Ultimatum Game, 49
  utilizing existential spite, 132
counterdominant spite
  Brexit referendum, 155–159
  characterizing, 47–48
  Clinton's 2016 campaign
    strategy, 150–153
  dominant spite and, 82–86
  freedom from domination, 123
  long-term social effects, 195
  online social media, 193–194
  politics triggering, 135
  as response to exploitation, 205
  societal advancement through,
    195
  suicide bombers, 169–170,
    183–184
  as a tool for justice, 190
counterelite, 160

creativity
    as benefit of spite, 6
    disagreeableness associated with,
        198–199
    existential spite and, 125–126,
        191
    stretch goals, 129–132
crisis negotiation, 29–30
Crockett, Molly, 88–89, 108, 154
Crome Yellow (Huxley), 193
cultural norms
    creation of gods, 164–165
    fairness and sharing, 53
    tradition outweighing reason,
        125
Cyberball (computer game),
    174–175

dangers of spiteful behavior, 35
dark factor of personality (D-factor),
    36
The Dark Knight (film), 139
Darwin, Charles, 43–44, 118, 179
Darwinian selection, 94–95
DaVita, 130–131
Dawkins, Richard, 27, 94, 97
de Quervain, Dominique, 55
death, 43
    Hamiltonian spite, 98–99
    infanticide, 27–30
    suicide bombers, 166–167
    through spousal spite, 26–29
deceit as tactic, 45
defining spite, 2–3
    Aristotle, 26
    evolved counterdominant
        and dominance-seeking
        behaviors, 47
    weak spite, 99
dehumanization, 61–62

delayed-benefit spite, 99
Delors, Jacques, 154–155
democracy
    in economic games, 186–187
    threatening elites, 201
Democratic National Convention,
    142–147
deservedness, 53–55
determinism, 121–122
deterrence: the purpose of
    punishment, 108
Di Tella, Rafael, 141
Dictator Game, 48–49, 51
    dominant spite, 83–86
    fair-minded people punishing
        unfairness, 76
    third-party costly punishment,
        64–67
disagreeableness, spite and,
    198–199
discord, age of, 159–160
disgust as a response to unfairness,
    56
do-gooder derogation, 72–75,
    105–107, 152–153
domestication, 43–44
dominance-seeking behaviors
    adaptations in, 45–46
    dominant individuals in
        hierarchies, 45
    evolution and, 39–41
    existential spite and, 122–124
    factors influencing, 46–47
    online attacks, 195–196
    as outcome of spite, 37
    pervasiveness of, 44–45
    purpose of punishment, 107
    reason as a form of, 123–124
    social-dominance orientation,
        180

dominant spite
  characterizing, 47–48
  competition boosting
    performance, 90–91
  counterdominant spite and,
    82–86
  effect of serotonin on, 88–91
  long-term advantages and costs,
    190–191
  political competition, 135–136
  religion as tool for, 165–166
  status competition, 80–82
dorsal striatum region, 89
dorsolateral prefrontal cortex,
  58–60, 94, 172–174
Dostoyevsky, Fyodor, 116–122
drivers of spiteful behavior, 4

eBay auctions, 15–17
economic games, 185–187
economic growth, unequal
  distribution of, 201
economic outcomes of Brexit, 158
economic theory
  explaining spite, 3
  Ultimatum Game, 19–21
economists, Ultimatum Game
  behavior in, 21
effortlessness of spite, 24–25
egalitarian behaviors and societies
  agricultural societies, 46–47
  dominant and counterdominant
    behaviors, 42–43
  evolution and, 39–42
  execution hypothesis, 44
  types of spiteful people, 37
elections
  Brexit referendum, 154–159
  counterdominant behaviors in
    2016, 150–153

elite betrayal, 141–142
equitable voting, 201–202
impact of spite on election
  results, 136–139
perception of fairness by liberal
  voters, 142–145
as a setting for spite, 135–136
spite-voters, 33–34
Trump's accusations of Clinton's
  untrustworthiness,
  148–150
US age of discord, 159–160
elite betrayal theory, 141
elite overproduction, 160
elite spiting: Brexit referendum,
  154–159
emotions
  controlling, 57–58
  importance of fairness to liberal
    voters, 142–145
  information generated from,
    58–59
  response to unfairness, 56–58
  social exclusion, 174–175
  of suicide bombers, 171–172
  use of costly punishment,
    104–105
  See also anger; moral emotions
empathy
  long-term role of spite, 190
  the need for chaos, 141
  rational and emotional brain
    activity, 58–59
  translating anger into spite,
    61–62
Enlightenment Now (Pinker), 123
Enlightenment principles
  the autonomous individual,
    111–113
  reason over tradition, 124

restricting freedom, 122–124
Russia's new men, 117–118
Ensslin, Gudrun, 12
environmental issues
identity fusion and, 185–187
stretch goals and, 131–132
envy motivating third-party
punishment, 65–66
Erdal, David, 42, 46–47
evolution
benefits of existential spite,
125–126
of costly punishment, 63–64
counterdominance- and
dominance-seeking sides, 46
dominance-seeking and
egalitarian tendencies,
39–42
execution hypothesis, 43–44
genetic effects on behavior,
93–95
genetic spite, 100–101
Hamiltonian spite, 96–98
motivations of suicide bombers,
168
reason and tradition, 124–125
evolutionary spite, 4–5
excellence as a benefit of spite, 6
execution hypothesis, 43–44
existential spite
benefits of, 125–126
boosting creativity, 191
drivers of, 122–124
long-term social effects of spite,
191
stretch goals, 129–133
the value of freedom, 122
existential threat, spite as, 35
Extinction Rebellion, 185
extreme parochial altruism, 180

fairness
Brexit referendum, 155–156
controlling anger-driven spite,
206–207
costly punishment, 50–51, 64
Dictator Game, 48–49
elements affecting perceptions
of, 51–54
generosity and spite, 37
negative assortment model,
102–103
punishment of unfairness,
55–56, 107
Trump's accusations of Clinton's
untrustworthiness, 148–150,
153
2016 Democratic primary
election, 142–148
Ultimatum Game, 18, 21
See also unfairness
faith and religion
gods meting out punishment,
163–165
permitting suicide bombing,
177–178
as tool for dominant spite,
165–166
Fallows, John, 63
family
benefits of spite for suicide
bombers, 168
children as weapons in spousal
spite, 27
counterdominant behaviors in
hunter-gatherer societies, 41
genetic effects on behavior,
93–96
identity fusion, 181–183
infanticide, 27–29
kin discrimination, 97

family (*continued*)
  reactance in children, 113
  Russia's generational split,
    117–118
fanaticism, 132–133
Farage, Nigel, 156–157
fascism, 127–128
Fathers and Sons (Turgenev),
  117–118
Father's Day, 27–29
fear, spite generating, 5
Fehr, Ernst, 49
Ferrari, Enzo, 30–31
Fiji: tradition and reason, 125
financial crisis (2008), 34, 201
Fincher, Katrina, 62
fire ants, 97–98
folklore, 1
Forber, Patrick, 101–103, 109
forgiveness, 204–206
Frazier, Kenneth, 131
free money, rejecting, 20–24. *See
  also* Ultimatum Game
free will, 113–115
freedom
  compulsion towards, 120–122
  counterdominant behavior
    relating to, 42–43
  driving existential spite, 122–124
  reactance, 113
  reason as a tool for liberation,
    116–117
functional spite, 99

Gaim, Medhanie, 129
game theory. *See* Ultimatum Game
generosity, 37, 72–75
genetic factors in behavior, 93–95,
  181–183
genetic spite, 96–99

Germany
  Baader-Meinhof group, 9–13,
    176, 191–193
  lasting shadows of war crimes,
    9–13
  Ultimatum Game, 16–19
Gilam, Gadi, 57–58
Ginges, Jeremy, 171–172
Gintis, Herbert, 49
gossip, 69–70
Gove, Michael, 123, 157
Greece, ancient
  autonomous individual concept,
    112–113
  defining spite, 26
  do-gooder derogation, 74
  long history of spite, 1
  spousal spite, 27
greenbeards, 97
grievances, suicide bombers
  harboring, 169–173, 178–179,
    183–185
Griggs, Rebecca, 130–131
group identity, 181–183
Guala, Francesco, 69–70
guilt, feelings of, 51
Güth, Werner, 17–19, 29–30

Habermas, Jürgen, 123
Hadza people (Tanzania)
  egalitarian tendencies, 41
  fairness and sharing, 53
  gossip as indirect punishment,
    69–70
  Ultimatum Game, 23
Hahn, Cordula, 26
Haidt, Jonathan, 42–43, 142, 152
Haldane, J.B.S., 95
Hamas, 184
Hamid, Nafees, 172

Hamilton, Alexander, 140
Hamilton, William, 94, 99
Hamiltonian spite, 96–98
harm, defining, 2–4
heart rate variability: controlling
    emotions, 57–58
hedonism, 127–128
Henrich, Joseph, 23, 124–125
Henry, Patrick, 113
heroes, 49
Herrmann, Benedikt, 73, 75
hierarchical organization
    dominance-seeking and
        counterdominant behaviors,
        46–47
    dominant and counterdominant
        spite, 47–48
    dominant individuals in
        hierarchies, 45
    evolution of egalitarianism, 44
    religion as a tool of spite,
        165–166
    status seeking, 46
hierarchy-legitimizing myths,
    180
hijacking, 9–10
The Hitchhiker's Guide to the Galaxy
    (Adams), 128–129
Hitler, Adolph, 6, 127–128
Hoffman, Elizabeth, 21–22
Holocaust, 6, 9–13
Homo economicus, 3, 19–20, 30, 50,
    81, 101–102
Homo reciprocans, 50, 72, 80
Homo rivalis, 80–81, 84
human rights movement, 195–196
hunter-gatherers
    counterdominant behavior, 85
    creation of gods in agricultural
        societies, 164–165

egalitarian tendencies, 40–41,
    46–47
Ultimate Game behaviors, 49
Huxley, Aldous, 193

identity fusion, 96, 181–183, 185
inclusive fitness, 95–96, 100–101,
    167–168
income distribution, 34
indirect punishment, 69–70
individualistic behavior, 81
inequality
    catalyzing the need for chaos,
        141
    dominant spite and restoring
        inequality, 83–86
    fairness in, 51–52
    human rights movement,
        195–196
    reproductive skew, 75
    wealth inequality influencing
        spite, 199–202
infanticide, 27–30
inhibitions, Ultimatum Game
    lowering, 24
insanity
    spiteful behavior as, 29–30
    sublime madness, 132–133
insects, sterile, 97–98
intentions influencing spiteful
    behavior, 52–53
Inuit people (Canada), 41, 69–70
Israeli Defense Forces, 183
Italy, 30–31

Jensen, Keith, 54
Johnstone, Rufus, 100
Joy of Destruction game,
    84–87
Ju/'hoansi people, 41

justice
  brain activity in anticipation of, 55–56
  counterdominant behaviors in humans, 42
  empathy and, 61

Kahneman, Daniel, 20, 131, 206
Kerr, Steve, 129
Kerry, John, 34
Khalid Sheikh Mohammed, 169–170
Kimbrough, Erik, 16
kin discrimination, 97–98
!Kung people, 40–41
Kurdeity, 182–183
Kurdish Peshmerga, 182–183

labor supply: age of discord, 159–160
Lamborghini, Ferruccio, 30–31
language: counterdominant behavior, 43
LaPiere, Richard, 15
last-place aversion, 34, 79–80
Laurin, Kristin, 164–165
liberty, counterdominant behavior relating to, 42–43
Libyan revolutionary soldiers, 182
lottery game, 179–181
Luce, Edward, 201
Lufthansa flight 181, 9–10

macaques, 99–100
Machiavellianism, 36
Machiguenga people (Peru), 23
Makransky, John, 207
male-male relationships, 41
Marcus, David, 14–16
Markovitz, Alan, 25

Marxism, 9–10, 192–193
McAuliffe, Katherine, 64, 105
McCullough, Michael, 206
Medea, 1, 27
medial orbitofrontal cortex, 55–56
meditation, 204–205
Mein Kampf (Hitler), 127–128
Merck & Co., 131
Milgram, Stanley, 112
Mill, John Stuart, 201–202
misunderstanding spite, 6
moat metaphor, 32
Moby Dick (Melville), 12–13, 111–112
Möller, Irmgard, 12
money. *See* Dictator Game; Ultimatum Game
money-burning experiments, 53–55, 84, 199–200
monkeys, status seeking in, 46
moral emotions
  counterdominant behaviors, 42–43
  Dictator Game, 48–49
  fair-minded people punishing unfairness, 76
  moral outrage, 56–57, 197–198
moral foundations theory, 142, 145
murder
  counterdominant behaviors in hunter-gatherer societies, 41
  development of tools for, 43
  infanticide, 27–30
  suicide bombers, 166–167
Murray, Douglas, 195–196
Musk, Elon, 32

Nachgeborenen ("later-born"), 9–10
Nama people (Namibia), 87
narcissism, 36

naturalism, 118, 120
Nazism, 127–128
need for chaos, 135, 139–141,
   191–192
negative assortment model, 101–104
negative indirect reciprocity, 99–101
negative relatedness, 98
neurostimulation, 59–60
Niebuhr, Reinhhold, 77, 132–133
Nietzsche, Friedrich, 103, 126–127,
   165–166, 190, 205–206
9/11 Commission Report, 185
9/11 terror attack, 169–170
Nineteen Eighty-Four (Orwell), 116
Nixon, Richard, 33–34
nonfinancial punishment, 71
nonsacred values, 172–175
Notes from the Underground
   (Dostoyevsky), 116–122
nuclear weapons, 35–36

Obama, Barack, 149–150, 201
obedience, 112–113
occasional spite, 100–101
Olympic Games, 11
On the Genealogy of Morality
   (Nietzsche), 103
On the Origin of Species (Darwin),
   118
Opel Automobile, 129
Orwell, George, 116, 127–128
Osborne, George, 155
outrage
   counterdominant spite in the
      2016 election, 151
   as response to unfairness, 56–57
   See also anger

parochial altruism, 179–181
Parrondo's paradox, 104

pecking order, 44–45
perceptual dehumanization, 62
personal fitness, 4
personality traits, 36–37
Petersen, Michael Bang, 139–141
philanthropic behavior, economists
   failing to engage in, 21
Pinker, Steven, 112, 123
Pleistocene, 40
politics
   Brexit referendum, 154–159
   counterdominant behaviors in
      humans, 42
   counterdominant behaviors in
      the 2016 general election,
      150–153
   dominant and counterdominant
      spite, 47–48
   egalitarian and dominance-
      seeking tendencies, 39–41
   elections as a setting for spite,
      135–136
   elite betrayal, 141
   ending terror campaigns,
      183–185
   fairness in the 2016 Democratic
      primary, 142–148
   global rise of populism, 153–155
   impact of spite on election
      results, 136–139
   Marxism and spite, 192–193
   moral foundations theory, 142
   the need for chaos, 139–141
   populism, 141–142, 153–155, 198
   social-dominance orientation,
      181
   spite in income distribution, 34
   spiteful voting, 33–34
Popular Front for the Liberation of
   Palestine, 9–10

populism, 141–142, 153–155, 198
positional bias, 87
power-seeking behavior, 39–41, 43
prefrontal cortex, 55–56, 58–60
Pretus, Clara, 174
primates
    counterdominant behaviors, 42
    deservedness, 53–55
    lack of spiteful behavior in, 20
    return-benefit spite, 99–100
    status seeking, 46
property damage through spite, 26
prosocial behavior, 49, 51, 81
psychological spite, 4, 100–101
psychopathy, 36
Pulp Fiction (film), 163–164
punishers, 104
punishment
    cheap spite, 70–71
    counterdominant and dominant
        spite, 82–86
    dehumanization of social norm-
        violators, 61–62
    effect of serotonin levels on spite,
        89–91
    gods meting out, 164–166
    moral outrage and, 197–198
    negative assortment model,
        102–104
    preventing suicide bombing,
        183–184
    societal benefits of costly
        punishment, 62–63
    Trump's accusations of Clinton's
        untrustworthiness, 148–150
    2016 elections, 136–138,
        146–147
    undeserved gains, 199–202
    of unfairness, 55–56
    See also costly punishment

quantifying spite, 14–15
queue jumping, 63–64

racism
    Clinton's accusations of
        Republican voters, 151–152
    dominant spite, 48
    racist talk versus racist behavior,
        15–16
    social exclusion and, 175
Raihani, Nichola, 105
Ram Dass, 111
Raspe, Jan-Carl, 11
The Rational Optimist (Ridley), 37
Rawls, John, 6
reactance, 113, 126
reason and rational behavior
    Constrained and Unconstrained
        Vision, 119–120, 124
    controlling anger-driven spite,
        203
    cost-benefit analysis, 58–60
    existential spite, 123–124
    inferiority to tradition, 124–125
    long-term social effects of spite,
        191
    maximizing happiness, 118–119
    obedience and autonomy,
        111–113
    social dominance, 45
    as a tool for liberation, 116–117
    tradition and, 124–126
recession (2008), 34, 201
reciprocity, 49–50
Reiss, J. Philip, 16
religion. *See* faith and religion
reproduction
    dominance-seeking in
        hierarchies, 45
    effect of serotonin levels on, 162

Hamiltonian spite, 96–97
measuring success, 4
reproductive skew, 75
reputation. *See* status seeking
resource competition
  competition motivating
    dominant spite, 87
  counterdominant behaviors in,
    41–42
  Dictator Game, 48–49
  dominance hierarchies and
    egalitarianism, 44–45
  effect of serotonin on, 88–91
  egalitarian agricultural societies,
    47
  long-term role of spite, 190
ressentiment, 205–206
retaliation
  fear of retaliation for spite,
    84–85
  online spite, 194
  preventing suicide bombing,
    183–184
  punishing unfairness without,
    68–69
  purpose of cooperation, 105–110
  purpose of punishment, 107–109
return-benefit spite, 99
Reuter, John, 178
rewarding spite, 193–195
Ridley, Matt, 37
risk-taking and risk aversion, 2–3,
  158–159
Ronson, Jon, 195
Rotemberg, Julio, 141
Rousseau, Jean-Jacques, 111–112

sacred values, 172–175, 184–185
sacrifice, 172
Sanders, Bernie, 143–149

Sartre, Jean-Paul, 11–12
Saving Private Ryan (film), 206
Schleyer, Hanns-Martin, 10–12
Schooler, Jonathan, 114
second-party costly punishment,
  67–68
The Secret of Our Success (Henrich),
  124–125
See's Candies, 32
selective serotonin reuptake
  inhibitors (SSRI), 161–162
self-control, exercising and
  maintaining, 24–25
self-destruction, 35–36
self-determination theory, 112–113
self-domestication, 43–44
self-harm, spiteful, 29–30
self-interest
  Brexit, 158–159
  defining spite, 3
  D-factor justifying spiteful
    behavior, 36–37
  environmental challenges and,
    185–187
  generosity and, 37
  incentivizing profit-maximizing
    businesses, 206–207
  individualistic behavior, 81–82
  in populists, 153–155
  suppressing, 60–61
  tragedy of the commons, 50–51
  Ultimatum Game behavior
    options, 18–19
  voting behavior, 33–34
The Selfish Gene (Dawkins), 27
selfishness
  cost-benefit analysis of spite,
    58–60
  counterdominant and dominant
    spite, 83

selfishness (*continued*)
  defining spite, 2–3
  effects of natural selection,
    94–96
  evolution of spite, 4
  guilt and, 51–53
  return-benefit spite, 99–100
Seneca, 202
serotonin levels, 161–162, 190
sexism, dominant spite and, 48
sexual behavior, 25, 30, 99–100
Shanabarger, Amy, 27–29
Shanabarger, Ronald, 27–29
Shanabarger, Tyler, 27–29
sharing
  cultural differences, 53
  deservedness, 53–55
  Dictator Game, 48–49
  fairness in behavior, 18
  identity fusion, 181–183
shattered assumptions, theory of,
  176–178
Sitkin, Sim, 130–131
slave morality, 165–166
SMART goals, 126
Smead, Rory, 101–103, 109
Smith, Adam, 3
So You've Been Publicly Shamed
  (Ronson), 195
social dominance, 44–45
social exclusion, 173–175
social information, 45
social justice movement,
  196–197
social marginalization, 141
social media platforms, 193–197
social norms and controls
  dehumanization of violators,
    61–62
  egalitarian societies, 40–41

indirect punishment for
    violating, 69–70
  third-party costly punishment,
    64, 192–194, 197
  tragedy of the commons, 50–51
social value orientation, 80–81
social-dominance orientation,
  180–181
socialism, 127–128
Somalia, 11
Southwest Airlines, 130
Sowell, Thomas, 119–120, 124
Speckhard, Anne, 170, 177–178
spite-voters, 33–34, 147, 201
spousal spite, 25–29
Stammheim Prison, Stuttgart,
  Germany, 11
status seeking
  competition motivating
    dominant spite, 86–87
  cost-benefit analysis of
    cooperation, 59–60
  dominance hierarchies, 45–46
  dominance-seeking behavior
    and counterdominant spite,
    195–196
  effect of testosterone levels on,
    90
  gossip affecting reputations, 70
  importance of reputations, 59–60
  last-place aversion, 79–80
  the need for chaos, 140–141
  the purpose of costly
    punishment, 109
  reputational benefits of spite, 101
  social justice movements,
    196–197
  suicide bombers, 167–168,
    183–184
  third-party spite, 67

stretch goals, 126–127, 129–133, 191

strong reciprocity, theory of, 49–50

Stroop task, 24–25

structural-demographic theory, 159–160

stumptail macaques, 99–100

Sturgeon, Nicola, 154

sublime madness, 132–133

suicide

Baader-Meinhof group deaths, 12–13

out of spite, 29–30

suicide bombers, 3–4, 166–171, 176–179, 183–185

superdelegates, 143–144

Sznycer, Daniel, 35

talking versus action, 13–16

Tanzania, 41

taxation: spite in income distribution, 34–35

terrorist acts

Baader-Meinhof group, 9–10

hijacking, 9–10

9/11 attack, 169–170

prevention strategies, 184–185

suicide bombers, 3–4, 166–171, 176–179, 183–185

testosterone, 90

Tetlock, Philip, 62

third-party costly punishment, 64–68, 192–194, 197

Thunberg, Greta, 185

tools, counterdominant behavior and, 43

Topless Prophet: The True Story of America's Most Successful Gentleman's Club Entrepreneur (Markovitz), 25

totalitarian states, 116

tradition, reason and, 124–126

tragedy of the commons, 50–51

transcranial direct-current stimulation, 57–58

Trump, Donald, 136–139, 141, 144, 147, 151, 160, 198

trust, 59–60

trustworthiness, 68

tryptophan, 88

Tullock, Gordon, 3

Turchin, Peter, 159–160

Turgenev, Ivan, 117–118

Tversky, Amos, 131

twins

genetic influences on behavior, 95–96

identity fusion, 181–182

Ultimatum Game behavior, 93–94

tyranny of the underdogs, 44, 165–166

Ultimatum Game

anger over unfair offers, 56–58

cheap punishment options, 70–71

cognitive reflection and, 203–204

controlling anger, 202

cost-benefit analysis, 58–60

costly punishment, 50, 64, 109

Dictator Game and, 48–49

do-gooder derogation, 72–75

economists' views of, 19–21

fair-minded people punishing unfairness, 76

fairness and generosity, 37, 53

genetic differences determining behavior, 93–95

Ultimatum Game (*continued*)
  human relationships mirroring,
    29–30
  mechanics and rules of, 18–19
  meditators' behavior, 204–205
  negative assortment, 101–103
  origins of, 13, 16–19
  populism as, 154
  the proposer's intent, 52–53
  punishing unfairness, 63, 68–69,
    76
  punishment for norm-violation,
    63
  the purpose of costly
    punishment, 109
  purpose of spiteful behavior,
    199
  serotonin and testosterone levels
    affecting choices, 88–90
  societal differences in behaviors,
    21–24
  status and dominant spite, 80–82
  welfare tradeoff ratio, 66–67
Unconstrained Vision, 119–120, 124
undeserved gains, 199–202
unfairness
  activating empathy, 61
  anger as response to, 56–58
  counterdominant and dominant
    spite, 83
  effect of antidepressants on
    perceptions, 161–162
  fair-minded people punishing,
    75–76
  gods punishing, 164–165
  moral outrage over, 197–198
  normalizing, 53
  politics as setting for, 135–136
  punishing without retaliation,
    68–69

purposes of, 105–110
  self-interest outweighing,
    59–60
  societal benefits of costly
    punishment, 62–63
  suicide bombers' perception of,
    169–173, 177
  suppressing self-interest, 60–61
  third-party costly punishment,
    65–66
  undeserved gains, 199–202
  See also fairness
unforced force, 123
United Kingdom: Brexit
  referendum, 123, 154–159
United Kingdom Independence
  Party (UKIP), 156
upward mobility, 200–201
utilitarianism, 118, 120, 127

values
  the autonomous individual, 113
  motivating suicide bombers,
    171–172
  sacred and nonsacred values,
    172–175, 184–185
van Lange, Paul, 80–81
vegetarianism, 74, 152–153
vengeance
  forgiveness and, 205–206
  sacred sources of, 163–164, 166
  suicide bombers normalizing,
    177
ventromedial prefrontal cortex,
  55–56, 58
virtue signaling, 196–197
virtues of vice, 6–7
Vohs, Kathleen, 113–114
Volkswagen, 129–130
voting, 33–34

wasps, Hamiltonian spite in, 98–99

Watson, James, 198

weak reciprocity, 49–50

weak spite, 84, 99

weapons, counterdominant behavior and, 43

Weinstein, Bret, 196

Weinstein, Eric, 198

WEIRD (Western, educated, industrialized, rich, and democratic) societies, 22–23

welfare tradeoff ratio, 66–67

What Happened (Clinton), 151–152

What Is to Be Done? (Chernyshevsky), 118–119

Whitehouse, Harvey, 182

WikiLeaks, 144, 149

Wilson, E.O., 96

Wilsonian spite, 96, 99

win-win situations: taxation for shifting inequalities, 34–35

women
   abused women's fear of spite, 26
   suicide bombers, 170–171, 177
   treatment in "egalitarian" societies, 41

Woolf, Virginia, 124

Wootton, Barbara, 126

Wrangham, Richard, 41, 43–44, 165–166

wrathful compassion, 207

written punishment, 70–71

Wynette, Tammy, 151

Simon McCarthy-Jones is associate professor of psychology at Trinity College Dublin. His research has appeared in *Nature Communications*, *Clinical Psychology Review*, and elsewhere. He has been featured in *Newsweek* and *Scientific American*, and on BBC News, ABC Radio, and the World Service. He lives in Ireland.